Dictionary of Environmental and Climate Change Law

Edited by

Nicholas A. Robinson

Pace University School of Law, USA

Wang Xi

Shanghai Jiao Tong University, China

Lin Harmon

Pace University School of Law, USA

Sarah Wegmueller

Pace University School of Law, USA

Edward Elgar
Cheltenham, UK • Northampton, MA, USA

Published by
Edward Elgar Publishing Limited
The Lypiatts
15 Lansdown Road
Cheltenham
Glos GL50 2JA
UK

Edward Elgar Publishing, Inc.
William Pratt House
9 Dewey Court
Northampton
Massachusetts 01060
USA

A catalogue record for this book
is available from the British Library

Library of Congress Control Number: 2012948843

This book is available electronically in the ElgarOnline.com
Law Subject Collection, E-ISBN 978 0 85793 578 6

ISBN 978 0 85793 577 9 (cased)

Typeset by Servis Filmsetting Ltd, Stockport, Cheshire
Printed by MPG PRINTGROUP, UK

CONTENTS

ACKNOWLEDGEMENTS

This dictionary is the product of a two-year research program by scholars and post-graduate students studying environmental law at Pace University School of Law in White Plains, New York. It was carried out under the auspices of the Pace University School of Law Center for Environmental Legal Studies with the cooperation of the Environmental and Resources Law Institute (ERLI) of the Faculty of Law at Shanghai Jiao Tong University. The following individuals contributed research time, working in Mandarin Chinese (or Standard Chinese, Putonghua, the official national language of China) and American English, to identify, refine, and edit the standard terms included in this book: Feng Jingru, Hou Xuemei, Laura Jensen, Audrey Kang, Sen Liang, Liu Zheyuan, Melissa Snyder, Sun Chih-yao, Nick Wagner, Yiyi Wang, Luo Wenjun, Wu Qi, Xue Long, Chen Yitong, and Sarah Wegmueller

A research team led by Professor Wang Xi identified a number of environmental law terms, in Chinese and without definitions, for this project. In addition, the team conducted a thorough review and correction on the first draft of the dictionary. The team members who contributed substantively to these terms are: Luo Wenjun, Xu Fengguo, Gu Dejin, Wang Xiaogang, Zhou Wei, Zhao Jun, Tang Tang, Deng Yang, Chen Xiaoli, and Zhu Dajun.

In addition, the editors are pleased to acknowledge the administrative support of Alexandra Dunn, Judith Weinstock, Karen Ferro, and Lorraine Rubich. The project has benefited from the many professional contributions made by the excellent Law Library staff at Pace University School of Law, including Professor Marie Stefanini Newman, Director of the Law Library, and John McNeill and Cynthia Pittson. The innovative studies of Professor John Nolon and the staff of the Land Use Law Center of Pace University School of Law have contributed to the inclusion of terminology of local government environmental and land-use law terms. Finally, the cooperation of the Pace University School of Law's Institute for Sino-American Comparative Environmental Law and its Director, Professor Richard L. Ottinger, and the Confucius Institute at Pace University, and its Director, Dr Niu Weihua, are gratefully acknowledged.

The editors acknowledge the permission granted by the Government Institutes Press and the Environmental Law Institute to include selected terms derived from their publications:
ISO 14000 Understanding the Environmental Standards, W.M. Von Zharen, © 1996 (Government Institutes Press).

WELL GROUNDED Using Local Land Use Authority to Achieve Smart Growth, John Nolon, © 2001.

While acknowledging with thanks the many contributors to this study, the editors at Pace University School of Law also acknowledge their responsibility for any errors or omissions, and invite users of this dictionary to contact them at:

Center for Environmental Legal Studies, Pace University School of Law
78 North Broadway, White Plains, New York 10603 www.law.pace.edu

INTRODUCTION AND USER'S GUIDE

The Need for this Dictionary

Environmental law remains a young and still maturing field of law, at international, national, and local levels. Its emergence with distinct legal concepts and terms dates from the 1970s, and two of the editors, Professor Nicholas A. Robinson and Professor Wang Xi, began their research and teaching of environmental law as the field emerged for the first time. Since those early days, a plethora of treaties, declarations, statutes, regulations, and policies for best management practices have been adopted to advance environmental protection and sustainable development. Most of these enactments are grounded in the environmental sciences, and attempt to align socio-economic practices with maintaining the quality of the ambient environment, which supports those practices. Nonetheless, there are often instances of inconsistent usage of terminology among legal enactments and across levels of government.

Both the United States of America and the Peoples Republic of China are developing environmental law rapidly, in response to perceived environmental problems. New legal provisions are a part of governmental agency responsibilities at the national level, state or provincial level, and the local level. Both nations have entered into a number of the same multilateral environmental agreements, other international treaties, and bilateral agreements. These treaties require implementation at state, provincial, or local levels. Both nations encourage investment in manufacturing and commercial trade by companies, which establish environmental management systems and norms for best practices to comply with environmental laws. Both nations recognize the terminology and programs of international organizations such as the United Nations Environment Programme (UNEP), the International Union for the Conservation of Nature and Natural Resources (IUCN), and the International Organization for Standardization (ISO).

As lawyers from China and the United States engage in trade negotiations, as their diplomats explore cooperation within rapidly evolving environmental treaty systems, or as administrators seek to implement environmental legal responsibilities, the need emerges for a common understanding of the meanings used for the same terms. Environmental law is complex, and often parties need efficient access to a source providing an understanding of its concepts and terms. The meaning of key Chinese concepts, such as the "circular economy," is not immediately perceived by Americans, and American legal terms, such as a "large concentrated animal feeding operation" (CAFO), need to be explained to Chinese speakers. In light of the complexities of environmental law, there are times when both Chinese and US legal personnel may not understand fundamental technological terms employed in specialized environmental regimes, such as "catalytic converters." Such technical terms arise in trade for products such as automobiles or in controls of atmospheric pollution. Basic terms taken from ecology or other fields of the environmental sciences and incorporated into a legal regime may be unfamiliar to legal experts who have not studied the environmental sciences. Most aspects of legal practices today have environmental law dimensions.

As China encourages international companies to build manufacturing facilities in the PRC, these enterprises need to understand Chinese legal terms, and the Chinese companies that are contracted to provide goods and services for the foreign enterprises need to understand what environmental requirements exist for the "supply chain" of these goods and services. For example, in February 2012, Apple announced that it had engaged an independent contractor to conduct third-party audits of its factories in China, such as those in Chengdu or in Shenzhen at Foxconn City, where iPhones, iPads and other Apple products are built. Apple's audits are to be released to the public. The generic terms used in these types of corporate social responsibility audits are widely understood in international practice but need translation. What this dictionary identifies is the Mandarin Chinese analogue for these terms as used in English. This dictionary is analogous to a linguistic road map for exploring the meaning in Mandarin of these terms.

Law schools in China are rapidly expanding their instruction in environmental law, and law schools in the United States and elsewhere are rapidly expanding their instruction in Chinese law. Both nations' law schools teach international environmental law as a growing subject within the field of public international law. The reference libraries for these studies have lacked a common reference providing terminology for environmental law in Mandarin Chinese and English.

In August 2011, more than 300 Chinese law professors, lawyers, and research scholars and some scholars from other countries met at the Guilin Science and Technology University to establish the China Association of Environment and Resource Law (CAERL) as an autonomous scholarly and professional society and to study issues in environmental law. The importance of a having a shared understanding of environmental law terminology in Mandarin and in English (and other languages, of course) is illustrated by noting some of the themes addressed by speakers at the symposium, these included ecological safety, climate change, energy and law, pollution from coal mines, the construction of new factories, food waste, and environmental risk. One of the editors of this dictionary, Sarah Wegmueller, addressed the Symposium on behalf of the Pace University School of Law Center for Environmental Legal Studies, and presented the remarks of Pace Dean Emeritus Richard L. Ottinger and co-editor Professor Nicholas A. Robinson in Mandarin.

Both the Center for Environmental Legal Studies of Pace University School of Law and the Environmental Resources Law Institute of Shanghai Jiao Tong University are members of the IUCN Academy of Environmental Law (www.iucnael.org) and meet annually in the Academy's Colloquium. The Inauguration Ceremony and the first Colloquium were held at Shanghai Jiao Tong University in November 2003, with the two editors of this dictionary, Professor Nicolas Robinson and Professor Wang Xi, as co-chairs. Now, the Academy has become a well-established and well-known non-governmental organization with membership from more than 150 universities all over the world. Many of its member law schools come from China and the United States. While the CAERL Symposium in China was conducted only in Mandarin, and comparable meetings such as those of the Environmental Law Section of the Association of American Law Schools (AALS) are conducted only in English, participants from China or the United States who do attend each others' meetings will benefit from having reference to a dictionary when the scholarly discussions become technical.

Similarly, the growing exchange of papers and research between American and Chinese scholars depends upon sharing a common vocabulary for environmental law. This dictionary serves an equally important service by providing a linguistic reference for interpreters at diplomatic, academic, or commercial meetings. The technical language of environmental law is foreign to many simultaneous interpreters, and there is a need for a reference to assist them in their important role. The same is true for those translating from written English to Mandarin, and the reverse. This dictionary is a first attempt to serve this communication need.

Comparable legal developments exist in the judiciary, where environmental law enforcement is of importance in both nations. US courts have produced an extensive body of decisions about environmental law, and the volume of these decisions continues to grow. A comparable trend is emerging in China. This dictionary will be helpful for the judiciary, enterprises, scholars, and others to understand each other's judicial decisions.

How the Dictionary was Constructed
No comparative environmental law dictionary as yet exists for Chinese–American environmental law. This book represents a beginning, but it will doubtless not be the last effort. To assemble this first compilation

of terms, the editors turned to several sources of law. First, colleagues at Shanghai Jiao Tong University compiled a set of terms in Mandarin for which no ready definitions existed. From this initial set, many terms were researched and definitions provided. Second, terms of environmental law that are defined in the international agreements ratified by both China and the United States were accessed and restated here. Third, where China has enacted environmental regulations that reflect standard practices with respect to technological or environmental issues, the official definitions employed in a comparable setting by the US Environmental Protection Agency and by other US agencies were drawn upon. Fourth, the terms normally used by UNEP or by expert bodies, such as the ISO, were accessed and drawn upon.

In all instances, this dictionary provides the appropriate Chinese Mandarin word and the Pinyin anglicized variant of the term. This facility in Mandarin will permit users to access wider meanings of terms from among Chinese legal authorities and references. Use can be made of the definitions in bilateral commercial and other dealings, whether by Chinese or by American parties and their lawyers. The terms can be accessed by diplomats elaborating existing treaty obligations or seeking to come to agreement on new legal obligations in international environmental negotiations. These terms, of course, may be used by any English-speaking parties with respect to China's usage, although standard British English usage will require different spellings than those employed here in American English usage,[1] as well as some variations in grammar.

Appended to this dictionary is a summary of short citations used in the definitions to indicate sources that may be accessed for further study. Following this summary is a roster of full citations to legal texts and authorities. These references will make it easier for users of this dictionary to undertake further research when reading a closer analysis of the terms and their uses. There is also a roster of selected Internet references in American usage, and Chinese users will find use of these references helpful in facilitating their further study of the terms.

Conclusion

In coming decades, it is likely that there will be a congruence of environmental laws across all nations, to reflect the reality that all States exist in

---

[1] The editors have retained the original British spellings of such official organizations as the International Labour Organisation and the United Nations Environment Programme, and of definitions drawn from original sources in British English, while using American English for all other purposes.

the same biosphere. Ultimately, Earth is a "closed" natural system, once the sun's energy has been received. The rapid changes that emission of greenhouse gases are producing in the Earth's climate, its weather patterns, the melting of glacial ice, and the rise of coastal sea levels, will invariably accelerate the adoption of common environmental laws to produce resilience at the local government level in order to adapt to, and recover from, environmental disruptions. Such forthcoming law reform will be more efficiently arrived at if all States use the same terminology. Many as yet undefined environmental concepts also will need to be invented, and will be new to all languages. Future dictionaries will build on the definitions included here. Such new steps may be facilitated by standardization of environmental terms. We encourage others to participate in the task of creating a shared global terminology in environmental law.

The United Nations Commission on Environment and Development, in its report entitled "Our common future,"[2] observed that "the Earth is one but the world is not."[3] Today, the rapid global elaboration of environmental laws allows States to speak with one voice. The definitions of terms used for establishing and sustaining ecological civilization contribute increasingly to that one voice. The editors dedicate this dictionary to the cause of more effective global cooperation for environmental stewardship, and to the orderly and peaceful settlement of environmental disputes.

---

[2]   Our common future, Oxford University Press, 1987 (also known as the "Bruntland" report, named after the Chair of the Commission, Norway's then Prime Minister Gro Harlem Bruntland).

[3]   Ibid., p. 27.

A

A Horizon (A 层, 腐殖堆积层) (A céng, fǔ zhí duī jī céng)

The mineral horizon at or near the surface in which an accumulation of humidified organic matter is mixed with the mineral material. Also, a plowed surface horizon. [Urban Soil Primer, p. 72 (2005)]

Absorbed Fraction (吸收份额) (xī shōu fèn é)

The fraction of energy emitted as a specified radiation type in a specified source region that is absorbed in a specified target tissue. [IAEA Safety Glossary, p. 11 (2007)]

Acceptance Criteria (验收标准) (yàn shōu biāo zhǔn)

Specified bounds on the value of a functional indicator or condition indicator used to assess the ability of a structure, system or component to perform its design function. [IAEA Safety Glossary, p. 12 (2007)]

Accessory Apartment (附属房) (fù shǔ fáng)

An accessory apartment is a second residential unit that may be contained within an existing single-family home, garage, or carriage house. An accessory apartment is usually required to be a complete housekeeping unit that can function independently with separate access, kitchen, bedroom, and sanitary facilities. [Nolon, Well Grounded, p. 445 (ELI 2001)]

Accessory Use (附属用途) (fù shǔ yòng tú)

An accessory use is the use of land that is subordinate, incidental to, and customarily found in connection with the principal use allowed on a lot by the zoning law. A garage is incidental to the principal use of a lot as a single-family residence and customarily found on a single-family parcel. [Nolon, Well Grounded, p. 445 (ELI 2001)]

Accident (意外事故) (yì wài shì gù)

Any unintended event, including operating errors, equipment failures and other mishaps, the consequences or potential consequences of which are not negligible from the point of view of protection or safety. [IAEA Safety Glossary, p. 12 (2007)]

Accident Precursor (事故前兆) (shì gù qián zhào)

An initiating event that could lead to accident conditions. [IAEA Safety Glossary, p. 13 (2007)]

Accreditation (资格认证) (zī gé rèn zhèng)

Procedure by which an authoritative body formally recognizes that a body or person is competent to carry out specific tasks. [Von Zharen, ISO 14000, p. 192 (1996)]

Accumulation (累积) (lěi jī)

The buildup or increase of a substance.

Acetic Acid (醋酸) (cù suān)

(1) An organic compound with the chemical formula CH_3CO_2H. (2) Acetic acid is found in all living organisms. Many people are familiar with acetic acid in its diluted form – vinegar. Pesticide products containing acetic acid are used in controlling a diverse group of weeds, including some grasses. It is readily broken down to carbon dioxide and water. Vinegar consists of approximately 5% acetic acid and 95% water. This is also the concentration of acetic acid when applied as a pesticide product. To be effective, acetic acid needs to contact the plant leaves; the acidity of the spray solution damages and dries out the leaves. [US EPA, Acetic Acid Factsheet (2001)]

Acid Aerosol (酸性气溶胶) (suān xìng qì róng jiāo)

Acidic liquid or solid particles that are small enough to become airborne. High concentrations of acid aerosols can be irritating to the lungs and have been associated with some respiratory disease. [US EPA Glossary of Indoor Air Quality Terms (2011)]

Action Level (行动阈值) (xíng dòng yù zhí)

A term used to identify the level of indoor radon at which remedial action is recommended. [US EPA Glossary of Indoor Air Quality Terms (2011)]

Actuated Equipment (驱动设备) (qū dòng shè bèi)

An assembly of prime movers and driven equipment used to accomplish one or more safety tasks. [IAEA Safety Glossary, p. 16 (2007)]

Actuation Device (驱动装置) (qū dòng zhuāng zhì)

A component that directly controls the motive power for actuated equipment. Examples of actuation devices include circuit breakers and relays that control the distribution and use of electric power and pilot valves controlling hydraulic or pneumatic fluids. [IAEA Safety Glossary, p. 16 (2007)]

Acute (急性的) (jí xìng de)

Of short duration (e.g., acute exposure). [US EPA, IRIS Glossary (2011)]

Adaptation (适应) (shì yìng)

Adjustment in natural or human systems to a new or changing environment. Adaptation to climate change refers to adjustment in natural or human systems in response to actual or expected climatic stimuli or their effects, which moderates harm or exploits beneficial opportunities. Various types of adaptation can be distinguished, including anticipatory and reactive adaptation, private and public adaptation, and autonomous and planned adaptation. [US EPA Glossary of Climate Change Terms (2011)]

Adaptation Fund (适应基金) (shì yìng jī jīn)

(1) A financial allocation, usually established by a national or international governmental authority, which is dedicated to providing support to redesign or reform activities or programs to align them with emerging or new environmental conditions, often used with respect to the effects of climate change. (2) An organization that finances projects and programs to help developing countries adapt to the negative effects of climate change. [Adaptation Fund, About the Adaptation Fund (2011)]

Additional Renewable Energy Tariff (可再生能源附加税) (kě zài shēng néng yuán fù jiā shuì)

A supplemental tax or fee, assessed by a governmental authority, as a surcharge to raise funds that are allocated to provide a subsidy for the installation of renewable sources for generating electricity, such as solar panels or wind turbines.

Additionality (额外性) (é wài xìng)

(1) Funding principle envisaged to ensure that the Global Environment Facility funds do not substitute for existing development finance but provide new and additional funding to produce agreed global environmental benefits. (2) Approval test for projects under the CDM [clean development mechanism] of the Kyoto Protocol. CDM projects are considered additional if they would not have taken place in the absence of the CDM. [Glossary of Terms for Negotiators of MEAs, p. 8 (2007)]. See **Clean Development Mechanism**.

Administrative Body (行政机构) (xíng zhèng jī gòu)

(1) An agency of government charged with the administration of specified laws. (2) Administrative bodies are created by local legislatures to undertake administrative functions such as the review of applications for

site plans, subdivisions, and special use permits. [Nolon, Well Grounded, p. 445 (ELI 2001)]

Administrative Or Legislative Measures (行政或立法措施) (xíng zhèng huò lì fǎ cuò shī)

Governmental rules or requirements established by an administrative agency for the promulgation of regulations, or by a legislative council or assembly for the enactment of statutes.

Adult Use (成人用品) (chéng rén yòng pǐn)

An adult use is a business that provides sexual entertainment or services to customers. Adult uses include: X-rated video shops and bookstores, live or video peep shows, topless or fully nude dancing establishments, combination book/video and "marital aid" stores, non-medical massage parlors, hot oil salons, nude modeling studios, hourly motels, body painting studios, swingers' clubs, X-rated movie theaters, escort service clubs, and combinations thereof. [Nolon, Well Grounded, p. 445 (ELI 2001)]

Advection (水平对流) (shuǐ píng duì liú)

The movement of a substance or the transfer of heat by the motion of the gas (usually air) or liquid (usually water) in which it is present. Sometimes used with the more common meaning – transfer of heat by the horizontal motion of the air – but in IAEA publications is more often used in a more general sense, particularly in safety assessment, to describe the movement of a radionuclide due to the movement of the liquid in which it is dissolved or suspended. Usually contrasted with diffusion, where the radionuclide moves relative to the carrying medium. [IAEA Safety Glossary, p. 17 (2007)]

Adverse Effects of Climate Change (气候变化的不利影响) (qì hòu biàn huà de bú lì yǐng xiǎng)

Changes in the physical environment or biota resulting from climate change . . . [that] have significant deleterious effects on the composition, resilience, or productivity of natural and managed ecosystems or on the operation of socio-economic systems or on human health and welfare. [United Nations Framework Convention on Climate Change, Art. 1 (1992)]

Advisory Opinion (咨询意见) (zī xún yì jiàn)

An advisory opinion is a report by a local administrative body, which does not have the authority to issue permits or adopt laws and regulations, prepared for the consideration by a local body that does. [Nolon, Well Grounded, p. 445 (ELI 2001)]

Aerobic Treatment Unit (ATU) (需氧处理装置) (xū yǎng chǔ lǐ zhuāng zhì)

A mechanical wastewater treatment unit that provides secondary wastewater treatment for a single home, a cluster of homes, or a commercial establishment by mixing air (oxygen) and aerobic and facultative microbes with the wastewater. ATUs typically use a suspended growth process (such as activated sludge-extended aeration and batch reactors), a fixed-film process (similar to a trickling filter), or a combination of the two treatment processes. [US EPA Glossary of NPDES Terms (2004)]

Aerosol (气溶胶) (qì róng jiāo)

A collection of airborne solid or liquid particles, with a typical size between 0.01 and 10 micrometers (μm) and residing in the atmosphere for at least several hours. Aerosols may be of either natural or anthropogenic origin. Aerosols may influence climate in two ways: directly, through scattering and absorbing radiation, and indirectly, through acting as condensation nuclei for cloud formation or modifying the optical properties and lifetime of clouds. The term has also come to be associated, erroneously, with the propellant used in "aerosol sprays." [US EPA Glossary of Climate Change Terms (2011)]

Aesthetic Resources (美学资源) (měi xué zī yuán)

Natural resources, such as open vistas, woods, scenic viewsheds, and attractive man-made settings whose appearance is an important ingredient in the quality of life in a community. [Nolon, Well Grounded, p. 445 (ELI 2001)]

Affected Areas (受影响地区) (shòu yǐng xiǎng dì qū)

Arid, semi-arid, and/or dry sub-humid areas affected or threatened by desertification. [Convention to Combat Desertification, Art. 1 (1994)]

Affected Countries (受影响国家) (shòu yǐng xiǎng guó jiā)

Countries whose lands include, in whole or in part, affected areas. [Convention to Combat Desertification, Art. 1 (1994)]

Affordable Housing (经济适用房) (jīng jì shì yòng fáng)

Housing developed through some combination of zoning incentives, cost-effective construction techniques, and governmental subsidies that can be rented or purchased by households who cannot afford market rate housing in the community. [Nolon, Well Grounded, p. 445 (ELI 2001)]

Afforestation (造林) (zào lín)

Planting of new forests on lands that historically have not contained forests. [US EPA Glossary of Climate Change Terms (2011)]

AFO (动物饲养作业) (dòng wù sì yǎng zuò yè)

Animal feeding operation.
See **Concentrated Animal Feeding Operation**.

Ageing (老化) (lǎo huà)

General process in which characteristics of a structure, system, or component gradually change with time or use . . . The process of becoming out of date (i.e., obsolete) owing to the evolution of knowledge and technology and associated changes in codes and standards. . . . [IAEA Safety Glossary, p. 17 (2007)] American English: **aging**.

Ageing Degradation (老化降解) (lǎo huà jiàng jiě)

Ageing effects that could impair the ability of a structure, system, or component to function within its acceptance criteria. Examples include reduction in diameter due to wear of a rotating shaft, loss in material toughness due to radiation embrittlement or thermal ageing, and cracking of a material due to fatigue or stress corrosion cracking. [IAEA Safety Glossary, p. 18 (2007)] American English: **aging**.

Agency (机构) (jī gòu)

An agency, under the State Environmental Quality Review Act (SEQRA), is any state or local agency, including zoning boards of appeals, local legislatures, planning boards, and, under certain circumstances, even building inspectors, that make discretionary decisions that may affect the environment. These agencies are subject to SEQRA regulations whenever taking an action. [Nolon, Well Grounded, p. 445 (ELI 2001)]

Agenda 21 (21世纪议程) (21 shì jì yì chéng)

(1) The action plan of recommendations adopted by consensus in 1992 by the United Nations Conference on Environment and Development (UNCED), and subsequently by the UN General Assembly, for national governments to implement in order to attain sustainable development progressively. (2) A comprehensive plan of action to be taken globally, nationally, and locally by organizations of the United Nations System, Governments, and Major Groups in every area in which human activity impacts on the environment. It was adopted by more than 178 Governments at the United Nations Conference on Environment and Development (UNCED) held in Rio de Janeiro, Brazil, June 3–14, 1992. [United Nations, Agenda 21 (1992)]

Aggrieved Party (受害方) (shòu hài fāng)

(1) A person whose rights or interests are adversely affected by an administrative or judicial decision. (2) Only aggrieved parties may appeal a reviewing body or local legislature's land use decision to the courts. The decision must result in some demonstrable harm to the party that is different from the impact of the decision on the community as a whole. [Nolon, Well Grounded, p. 445 (ELI 2001)]

Agreement on the Conservation of Antarctic Fauna and Flora Measures (南极动植物保护措施协定) (nán jí dòng zhí wù bǎo hù cuò shī xié dìng)

Declared "no longer current" as of January 7, 2011, this Agreement designated the Antarctic as a place warranting special conservation. [Decision 1 (2011) - ATCM XXXIV - CEP XIV, Buenos Aires]

Agreement on the Network of Aquaculture Centers in Asia and the Pacific (亚太水产业中心网络协议) (yà tài shuǐ chǎn yè zhōng xīn wǎng luò xié yì)

A treaty signed by 35 countries in 1988 regarding aquaculture and fisheries. [Agreement on the Network of Aquaculture Centers in Asia and the Pacific (1988)]

Agricultural Land (农业用地) (nóng yè yòng dì)

(a) Land used for farming, agricultural, horticultural, viticultural, vegetable-growing, market gardening, pastoral, grazing, poultry farming, silvicultural, floricultural or piscicultural purposes, and (b) any other land declared to be farming lands for the purposes of soil legislation. [IUCN, Protected Area Management Categories (1994)]

Agricultural Land Protection (农业土地保护) (nóng yè tǔ dì bǎo hù)

Any law, regulation, board, or process that has as its objective the preservation of farming on land dedicated to agricultural use. Examples include agricultural zoning, farmland preservation boards, property tax relief for farmers, and anti-nuisance laws. [Nolon, Well Grounded, p. 445 (ELI 2001)]

Agricultural Zoning District (农业规划区) (nóng yè guī huà qū)

An agricultural zoning district is a designated portion of the municipality where agricultural uses are permitted as-of-right and non-farm land uses are either prohibited or allowed subject to limitations or conditions imposed to protect the business of agriculture. [Nolon, Well Grounded, p. 445 (ELI 2001)]

AHU (空气处理装置) (kōng qì chǔ lǐ zhuāng zhì)

See **Air Handling Unit**.

Aichi, Japan Rice Bran Oil Event (日本爱知县米糠油事件) (rì běn ài zhī xiàn mǐ kāng yóu shì jiàn)

In the *Yusho* outbreak, which occurred mostly in Fukuoka and Nagasaki prefectures in Japan in 1968, PCBs and very small amounts of dioxins were detected in the contaminated rice bran oil, and in the blood and adipose tissue of victims. The manifestations of *Yusho* included comedo, increased prominence of the pores, increased conjunctival secretion, pigment deposition in the skin, nail deformity and pigmentation, and chloracne. [Japan Envt'l Health Comm. Report, pp.6-7 (1999)]

Air (空气) (kōng qì)

The invisible, odorless, and tasteless mixture of gases which surrounds the Earth. [Massachusetts v. E.P.A., 549 U.S. 497, 559 (2007)]

Air Cleaning (空气净化) (kōng qì jìng huà)

An IAQ control strategy to remove various airborne particulates and/or gases from the air. The three types of air cleaning most commonly used are particulate filtration, electrostatic precipitation, and gas sorption. [US EPA Glossary of Indoor Air Quality Terms (2011)]

Aircraft (飞行器) (fēi xíng qì)

Cargo aircraft: any aircraft, other than a passenger aircraft, which is carrying goods or property . . .

Passenger aircraft: an aircraft that carries any person other than a crew member, a carrier's employee in an official capacity, an authorized representative of an appropriate national authority, or a person accompanying a consignment. [IAEA Safety Glossary, p. 19 (2007)]

Air Exchange Rate (空气置换率) (kōng qì zhì huàn lǜ)

The rate at which outside air replaces indoor air in a space. Expressed in one of two ways: the number of changes of outside air per unit of time air changes per hour (ACH); or the rate at which a volume of outside air enters per unit of time – cubic feet per minute (cfm). [US EPA Glossary of Indoor Air Quality Terms (2011)]

Air Handling Unit (AHU) (空气处理装置) (kōng qì chǔ lǐ zhuāng zhì)

For purposes of this document refers to equipment that includes a blower or fan, heating and/or cooling coils, and related equipment such as controls, condensate drain pans, and air filters. Does not include ductwork, registers or grilles, or boilers and chillers. [US EPA Glossary of Indoor Air Quality Terms (2011)]

Air Pollutant (空气污染物) (kōng qì wū rǎn wù)

Any air pollution agent or combination of such agents, including any physical, chemical, . . . substance or matter which is emitted into or otherwise enters the ambient air. [Massachusetts v. E.P.A., 549 U.S. at 506]

Air Pollution (大气污染) (dà qì wū rǎn)

The existence in the air of substances in concentrations that are determined unacceptable. Contaminants in the air we breathe come mainly from manufacturing industries, electric power plants, automobiles, buses, and trucks. [NASA Earth Observatory (2012)]

Air Toxics (毒性空气污染物) (dú xìng kōng qì wū rǎn wù)

Chemicals in the air that are known or suspected to cause cancer or other serious health effects, such as reproductive problems or birth defects. Air toxics are also known as "hazardous air pollutants." Mobile sources emit a number of air toxics associated with both long-term and short-term health effects in people, including heart problems, asthma symptoms, eye and lung irritation, cancer, and premature death. [US EPA Glossary of Climate Change Terms (2011)]

ALARA (合理可行尽量低原则) (hé lǐ kě xíng jìn liàng dī yuán zé)

Stands for as low as reasonably achievable. [IAEA Safety Glossary, p. 19 (2007)]

Albacore Tuna (长鳍金枪鱼) (cháng qí jīn qiāng yú)

Thunnus alalunga. A highly migratory species of fish. [Law of the Sea Convention, Annex 1 (1982)]

Albedo (反射率) (fǎn shè lǜ)

The fraction of solar radiation reflected by a surface or object, often expressed as a percentage. Snow covered surfaces have a high albedo; the albedo of soils ranges from high to low; vegetation covered surfaces and oceans have a low albedo. The Earth's albedo varies mainly through varying cloudiness, snow, ice, leaf area and land cover changes. [US EPA Glossary of Climate Change Terms (2011)]

Algae (藻类) (zǎo lèi)

Simple rootless plants that grow in sunlit waters in relative proportion to the amounts of nutrients available. They are food for fish and small aquatic animals, and a factor in eutrophication. [NASA Earth Observatory (2012)]

Algal Blooms (水华) (shuǐ huá)

Sudden spurts of algal growth due to greatly increased amounts of phosphorus entering the aquatic ecosystem from sewage systems and agricultural fertilizers. Excessive growth of the algae causes destruction of many of the higher links of the food web. Algae that die and sink to the bottom at the end of the growing season stimulate massive growth of bacteria the following year, resulting in depletion of oxygen in the deeper water layers. This may result in fish kills and replacement with less valuable species[,] . . . [which] may be more tolerant of increased phosphorus levels. Deoxygenation . . . may [also] cause chemical changes in the mud on the bottom, producing increased quantities of chemicals and toxic gases. All these changes further accelerate the eutrophication (aging) of the aquatic ecosystem. [NASA Earth Observatory (2012)]

Allergen (过敏原) (guò mǐn yuán)

A substance capable of causing an allergic reaction because of an individual's sensitivity to that substance. [US EPA Glossary of Indoor Air Quality Terms (2011)]

Allergic Rhinitis (过敏性鼻炎) (guò mǐn xìng bí yán)

Inflammation of the mucous membranes in the nose that is caused by an allergic reaction. [US EPA Glossary of Indoor Air Quality Terms (2011)]

Alpine Soil Treaty (阿尔卑斯土壤条约) (ā ěr bēi sī tǔ rǎng tiáo yuē)

See **Protocol on the Implementation of the Alpine Convention of 1991 in the Field of Soil Conservation**.

Alternative Energy (新能源) (xīn néng yuán)

Energy derived from nontraditional sources (e.g., compressed natural gas, solar, hydroelectric, wind). [US EPA Glossary of Climate Change Terms (2011)]

Aluminum Toxic Waste Spill In Hungary (匈牙利铝制有毒废物泄漏事件) (xiōng yá lì lǚ zhì yǒu dú fèi wù xiè lòu shì jiàn)

An environmental disaster in which 185 million gallons of caustic waste escaped from an aluminum waste reservoir in Hungary on October 4, 2010. [NASA, Toxic Sludge in Hungary (2012)]

Amazon Cooperation Treaty (亚马逊河合作条约) (yà mǎ xùn hé hé zuò tiáo yuē)

A treaty signed in 1978 by eight countries regarding sustainable development in the Amazon region. [Amazon Cooperation Treaty, July 3, 1978]

Amortization of Nonconforming Uses (不一致使用的分期摊销) (bú yī zhì shǐ yòng de fēn qī tān xiāo)

Nonconforming uses that are particularly inconsistent with zoning districts within which they exist and not immediately dangerous to public health or safety may be terminated or amortized within a prescribed number of years. This amortization period allows the landowner to recoup some or all of his investment in the offensive nonconforming use. [Nolon, Well Grounded, pp. 445–6 (ELI 2001)]

Analysis (分析) (fēn xī)

In general . . . analysis suggests the process and result of a study aimed at understanding the subject of the analysis. [IAEA Safety Glossary, p. 19 (2007)]

Ancillary Material (辅助材料) (fǔ zhù cái liào)

Material input that is used by the system producing the product, but is not used directly in the formation of the product. [Von Zharen, ISO 14000, p. 192 (1996)]

Animal Dander (动物皮屑) (dòng wù pí xiè)

Tiny scales of animal skin. [US EPA Glossary of Indoor Air Quality Terms (2011)]

Animal Feeding Operation (AFO) (动物饲养作业) (dòng wù sì yǎng zuò yè)

Lot or facility (other than an aquatic animal production facility) where the following conditions are met: animals (other than aquatic animals) have been, are, or will be stabled or confined and fed or maintained for a total of 45 days or more in any 12-month period, and crops, vegetation, forage growth, or post-harvest residues are not sustained in the normal growing season over any portion of the lot or facility. [US EPA Glossary of NPDES Terms (2004)]

Antarctic Treaty (南极条约) (nán jí tiáo yuē)

[T]he earliest of the post-World War II arms limitation agreements . . . It internationalized and demilitarized the Antarctic Continent and provided for its cooperative exploration and future use. Signed December 1, 1959 by 13 countries and entered into force June 23, 1961. [US DOJ, The Antarctic Treaty (2001)]

Antarctic Treaty Protocol on Environmental Protection (南极条约环境保护议定书) (nán jí tiáo yuē huán jìng bǎo hù yì dìng shū)

A statement of the environmental principles governing the conduct of States Parties in relation to the Antarctic continent. The Parties to the Protocol "commit themselves to the comprehensive protection of the Antarctic environment and dependent and associated ecosystems and hereby designate Antarctica as a natural reserve, devoted to peace and science." The Protocol aims to protect the Antarctic environment and dependent and associated ecosystems, and preserve the intrinsic value of Antarctica – its aesthetic values as well as its status as a place of research. This Protocol was adopted on October 3, 1991, in Madrid, Spain, and entered into force in 1998. [UNEP, Protocol to the Antarctic Treaty on Environmental Protection (1998)]

Antarctica Case (England v. Argentina, Chile) (ICJ, 1956) (南极洲案) (nán jí zhōu àn) (英国诉阿根廷、智利, 国际法院, 1956年)) (yīng guó sù ā gēn tíng, zhì lì, guó jì fǎ yuàn, 1956 nián)

A 1956 decision by the International Court of Justice regarding the sovereignty of certain islands and lands in the Antarctic. [Antarctica (1956)]

Anthropogenic (人为的) (rén wéi de)

Made by people or resulting from human activities. Usually used in the context of emissions that are produced as a result of human activities. [US EPA Glossary of Climate Change Terms (2011)]

Anticipated Transient Without Scram (ATWS) (未能紧急停堆的预期瞬态) (wèi néng jǐn jí tíng duī de yù qī shùn tài)

For a nuclear reactor, an accident for which the initiating event is an anticipated operational occurrence and in which the fast shutdown system of the reactor fails to function. [IAEA Safety Glossary, p. 21 (2007)]

Antidegradation Policy (防止水质退化政策) (fáng zhǐ shuǐ zhì tuì huàzhèng cè)

Policies which ensure protection of water quality for a particular water body where the water quality exceeds levels necessary to protect fish and wildlife propagation and recreation on and in the water. This also includes special protection of waters designated as outstanding natural resource waters. Antidegradation plans are adopted by each state to minimize adverse effects on water. [US EPA Glossary of NPDES Terms (2004)]

Antimicrobial (杀菌剂) (shā jūn jì)

Agent that . . . [prevents] microbial growth. [US EPA Glossary of Indoor Air Quality Terms (2011)]

Appellate Jurisdiction (上诉管辖) (shàng sù guǎn xiá)

A zoning board of appeals has appellate jurisdiction to review determinations of the zoning enforcement officer. Denials of building permits and determinations that proposed land uses do not meet the zoning law's standards may be appealed to the zoning board of appeals. Land-use decisions of the zoning board of appeals, planning board, and local legislature may be appealed to the courts, which exercise appellate jurisdiction over them. [Nolon, Well Grounded, p. 446 (ELI 2001)]

Applicant (申请人) (shēn qǐng rén)

A legal person who applies to a regulatory body for authorization to undertake specified activities. Strictly, an applicant would be such from the time at which an application is submitted until the requested authorization is either granted or refused. However, the term is often used a little more loosely than this, particularly in cases where the authorization process is long and complex. [IAEA Safety Glossary, p. 21 (2007)]

Approval (批准) (pī zhǔn)

An approval is a discretionary decision made by a local agency to issue a permit, certificate, license, lease, or other entitlement or to otherwise authorize a proposed project or activity. [Nolon, Well Grounded, p. 446 (ELI 2001)]

Approved Site or Facility (批准的场地或设施) (pī zhǔn de chǎng dì huò shè shī)

A site or facility for the disposal of hazardous wastes or other wastes which is authorized or permitted to operate for this purpose by a relevant authority of the State where the site or facility is located. [Basel Convention, Art. 2, No. 5 (1989)]

Arbitral Procedure (仲裁程序) (zhòng cái chéng xù)

Rules providing for the submission of claims to an arbitrator or a panel of arbitrators for a decision to resolve the claims.

Archipelago (群岛) (qún dǎo)

A group of islands, including parts of islands, interconnecting waters and other natural features, which are so closely interrelated that such islands, waters and other natural features form an intrinsic geographical, economic and political entity, or which historically have been regarded as such. [Law of the Sea Convention, Art. 46 (1982)]

Architectural Review Board (建筑审查委员会) (jiàn zhù shěn chá wěi yuán huì)

A body that reviews proposed developments for their architectural congruity with surrounding developments and either renders an advisory opinion on the matter or is authorized to issue or deny a permit. Its review is based upon design criteria or standards adopted by the local legislature. [Nolon, Well Grounded, p. 446 (ELI 2001)]

Area of Low Pest or Disease Prevalence (低虫害和疾病流行地区) (dī chóng hài hé jí bìng liú xíng dì qū)

An area, whether all of a country, part of a country, or all or parts of several countries, as identified by the competent authorities, in which a specific pest or disease occurs at low levels and which is subject to effective surveillance, control, or eradication measures. [Agreement on the Application of Sanitary and Phytosanitary Measures, p. 78 (2012)]

Area under the National Jurisdiction of a State
(在一国国家管辖下的区域) (zài yī guó guó jiā guǎn xiá xià de qū yù)

Any land, marine area or airspace within which a State exercises administrative and regulatory responsibility in accordance with international law in regard to the protection of human health or the environment. [Basel Convention, Art. 2, No. 9 (1989)]

Area Variance (地区差异) (dì qū chā yì)

This is a variance that allows for the use of land in a way that is not permitted by the dimensional or physical requirements of the zoning law. This type of variance is needed when a building application does not comply with the setback, height, lot, or area requirements of the zoning law. For example, if an owner wants to build an addition to a house that encroaches into the side-yard setback area, that owner must apply to the zoning board of appeals for an area variance. [Nolon, Well Grounded, p. 446 (ELI 2001)]

Arid, Semi-Arid and Dry Sub-Humid Areas
(干旱,半干旱和亚湿润干旱地区) (gān hàn, bàn gān hàn hé yà shī rùn gān hàn dì qū)

Areas, other than polar and sub-polar regions, in which the ratio of annual precipitation to potential evapotranspiration falls within the range from 0.05 to 0.65. [Convention to Combat Desertification, Art. 1 (1994)]

Arsenic (砷) (shēn)

(1) [A] semi-metal element in the periodic table. It is odorless and tasteless. It enters drinking water supplies from natural deposits in the earth or from agricultural and industrial practices. Non-cancer effects can include thickening and discoloration of the skin, stomach pain, nausea, vomiting, diarrhea, numbness in hands and feet, partial paralysis, and blindness. Arsenic has been linked to cancer of the bladder, lungs, skin, kidney, nasal passages, liver, and prostate. [US EPA, Arsenic in Drinking Water (2011)] (2) [F]or most people, food is the major source of exposure. Acute (short-term) high-level inhalation exposure to arsenic dust or fumes has resulted

in gastrointestinal effects (nausea, diarrhea, abdominal pain); central and peripheral nervous system disorders have occurred in workers acutely exposed to inorganic arsenic. Chronic (long-term) inhalation exposure to inorganic arsenic in humans is associated with irritation of the skin and mucous membranes. Chronic oral exposure has resulted in gastrointestinal effects, anemia, peripheral neuropathy, skin lesions, hyperpigmentation, and liver or kidney damage in humans. Inorganic arsenic exposure in humans, by the inhalation route, has been shown to be strongly associated with lung cancer, while ingestion of inorganic arsenic in humans has been linked to a form of skin cancer and also to bladder, liver, and lung cancer. EPA has classified inorganic arsenic as a Group A human carcinogen. [US EPA, Arsenic Compounds (2007)]

Article 78 Proceeding (第七十八条诉讼) (dì qī shí bā tiáo sù sòng)

An Article 78 Proceeding refers to an article of the Civil Practice Law and Rules that allows aggrieved persons to bring an action against a government body or officer. This device allows review of state and local administrative proceedings in court. [Nolon, Well Grounded, p. 446 (ELI 2001)]

As-of-Right (依法使用土地) (yī fǎ shǐ yòng tǔ dì)

An as-of-right use is a use of land that is permitted as a principal use in a zoning district. In a single-family district, the construction of a single-family home is an as-of-right use of the lot. [Nolon, Well Grounded, p. 446 (ELI 2001)]

Asbestos Case (Canada v. European Union) (WT/DS135, 1998) (加拿大诉欧盟石棉案) (WT/DS135, 1998) (jiā ná dà sù ōu méng shí mián àn) (WT/DS135, 1998))

A 1998 WTO case in which Canada challenged the complete ban of all uses of asbestos by France. In 2000, the dispute settlement panel held that the ban was justified to protect the health of French workers under the 1994 GATT, Art. XX (b).

ASEAN (Association of Southeast Asian Nations) (东南亚国家联盟) (dōng nán yà guó jiā lián méng)

(1) An intergovernmental regional organization of States situated in Southeast Asia, for cooperation on economic and social integration. (2) Established on 8 August 1967 in Bangkok, Thailand, with the signing of the ASEAN Declaration by Indonesia, Malaysia, Philippines, Singapore and Thailand. ASEAN aims to foster cooperation, peace, and economic growth in the region. [ASEAN, About ASEAN (2009)]

Assessment (评估) (píng gū)

An estimate or determination of the significance, importance, or value of something. [Von Zharen, ISO 14000, p. 192 (1996)]

Assessment System (评估制度) (píng gū zhì dù)

Procedural and managerial rules for conducting an assessment leading to issue of a certification document and its maintenance. [Von Zharen, ISO 14000, p. 192 (1996)]

Assisted (by the IAEA) Operation ((由国际原子能机构) 提供援助的行动) ((yóu guó jì yuán zǐ néng jī gòu) tí gōng yuán zhù de xíng dòng)

An operation undertaken by a State or group of States to which assistance is provided by or through the IAEA in the form of materials, services, equipment, facilities, or information pursuant to an agreement between the IAEA and that State or group of States. [IAEA Safety Glossary, p. 26 (2007)]

Associated Off-Shore Processing (相关近海加工) (xiāng guān jìn hǎi jiā gōng)

In manufacturing of products, the process that an enterprise arranges by contract to obtain services related to, or partial manufacturing of components, or assembly of a product, which are provided in a State or territory located outside the State where the primary manufacturing is situated. The reference to the location being off the shore or border of the State where manufacturing occurs applies to any place outside the borders, not only with respect to seashores.

Association of Southeast Asian Nations (东南亚国家联盟) (dōng nán yà guó jiā lián méng)

See **ASEAN**.

Atmosphere (大气层) (dà qì céng)

(1) The gaseous envelope surrounding the Earth. The dry atmosphere consists almost entirely of nitrogen (78.1% volume mixing ratio) and oxygen (20.9% volume mixing ratio), together with a number of trace gases, such as argon (0.93% volume mixing ratio), helium, radiatively active greenhouse gases such as carbon dioxide (0.035% volume mixing ratio), and ozone. In addition the atmosphere contains water vapor, whose amount is highly variable but typically 1% volume mixing ratio. The atmosphere also contains clouds and aerosols. [US EPA Glossary of Climate Change Terms (2011)] (2) The mass of air surrounding the earth and bound to it more or less permanently by the Earth's gravitational attraction. [NOAA, Terms Used By Meteorologists (2012)]

Atmospheric Lifetime (大气中的寿命) (dà qì zhōng de shòu mìng)

The lifetime of a greenhouse gas refers to the approximate amount of time it would take for the anthropogenic increment to an atmospheric pollutant concentration to return to its natural level (assuming emissions cease) as a result of either being converted to another chemical compound or being taken out of the atmosphere via a sink. This time depends on the pollutant's sources and sinks as well as its reactivity. The lifetime of a pollutant is often considered in conjunction with the mixing of pollutants in the atmosphere; a long lifetime will allow the pollutant to mix throughout the atmosphere. Average lifetimes can vary from about a week (sulfate aerosols) to more than a century (chlorofluorocarbons (CFCs), carbon dioxide). [US EPA Glossary of Climate Change Terms (2011)]

Attenuation 衰减) (shuāi jiǎn)

The reduction in intensity of radiation passing through matter due to processes such as absorption and scattering. By analogy, also used in other situations in which some radiological property, characteristic or parameter is gradually reduced in the course of passing through a medium (e.g., the reduction in activity concentration in groundwater passing through the geosphere due to processes such as sorption). [IAEA Safety Glossary, p. 27 (2007)]

ATWS (未能紧急停堆的预期瞬态) (wèi néng jǐn jí tíng duī de yù qī shùn tài)

See **Anticipated Transient Without Scram**.

Audit (审计) (shěn jì)

A planned, independent, and documented assessment to determine whether agreed-upon requirements are being met. [Von Zharen, ISO 14000, p. 193 (1996)]

Audit Conclusion (审计结论) (shěn jì jié lùn)

Professional judgment or opinion expressed by an auditor about the subject matter of the audit, based on and limited to reasoning the auditor has applied to audit findings. [Von Zharen, ISO 14000, p. 193 (1996)]

Audit Criteria (审计准则) (shěn jì zhǔn zé)

Policies, practices, procedures, or requirements against which the auditor compares collected evidence about the subject matter. [Von Zharen, ISO 14000, p. 193 (1996)]

Audit Findings (审计认定) (shěn jì rèn dìng)

Result of the evaluation of the collected audit evidence compared against the agreed audit criteria. [Von Zharen, ISO 14000, p. 193 (1996)]

Audit Program (审计程序) (shěn jì chéng xù)

The organizational structure, commitment and documented methods used to plan and perform audits. [Von Zharen, ISO 14000, p. 193 (1996)]

Audit Team (审计小组) (shěn jì xiǎo zǔ)

Group of auditors, or a single auditor designated to perform a given audit. The audit team may also include technical experts and auditors in training. One of the auditors on the audit team performs the function of lead auditor. [Von Zharen, ISO 14000, p. 193 (1996)]

Auditee (受审单位) (shòu shěn dān wèi)

Organization to be audited. [Von Zharen, ISO 14000, p. 193 (1996)]

Auditor (审计员) (shěn jì yuán)

Individual performing an environmental audit, or part thereof, who meets the criteria specified in ISO 14012. [Von Zharen, ISO 14000, p. 193 (1996)]

Australia-Salmon Case (WT/DS18) (澳大利亚鲑鱼案 (WT/DS18)) (ào dà lì yà guī yú àn (WT/DS18))

A dispute settlement between Canada and Australia to address Australia's prohibitions of salmon imports from Canada based on a quarantine regulation. In 1998, the WTO found that the Australian requirement requiring heat treatment of imported salmon violates international law rules. [Australia-Salmon Case (1998)]

Authorization (批准 (授权)) (pī zhǔn (shòu quán))

The granting by a regulatory body or other governmental body of written permission for an operator to perform specified activities. Authorization could include, for example, licensing, certification or registration. The term authorization is also sometimes used to describe the document granting such permission. Authorization is normally a more formal process than approval. [IAEA Safety Glossary, p. 27 (2007)]

Authorized Program or Authorized State
(获授权项目或获授权州) (huò shòu quán xiàng mù huò shòu quán zhōu)

A state, Territorial, Tribal, or interstate NPDES program which has been approved or authorized by EPA under 40 CFR Part 123. [US EPA Glossary of NPDES Terms (2004)]

Authorized Termination of Responsibility
(经批准的责任终止) (jīng pī zhǔn de zé rèn zhōng zhǐ)

The release by the regulatory body of an operator (or a former operator) from any further regulatory responsibilities in relation to an authorized facility or authorized activity. This may be a separate process from termination of an authorization, e.g. termination of the responsibility to maintain active institutional control over a repository. [IAEA Safety Glossary, p. 28 (2007)]

Authorized Transfer (获授权转让) (huò shòu quán zhuǎn ràng)

The transfer of regulatory responsibility for specified radioactive material from one operator to another. This does not necessarily involve any movement of the material itself. [IAEA Safety Glossary, p. 28 (2007)]

Availability ((时间)可得性) ((shǐjiān) kě dé xìng)

The fraction of time for which a system is capable of fulfilling its intended purpose. Reliability represents essentially the same information, but in a different form. [IAEA Safety Glossary, p. 28 (2007)]

Average Monthly Discharge Limitations (月平均排放量限度) (yuè píng jūn pái fàng liàng xiàn dù)

The highest allowable average of daily discharges over a calendar month, calculated as the sum of all daily discharges measured during that month divided by the number of days on which monitoring was performed (except in the case of fecal coliform). [US EPA Glossary of NPDES Terms (2004)]

Average Weekly Discharge Limitation (周平均排放量限度) (zhōu píng jūn pái fàng liàng xiàn dù)

The highest allowable average of daily discharges over a calendar week, calculated as the sum of all daily discharges measured during a calendar week divided by the number of daily discharges measured during that week. [US EPA Glossary of NPDES Terms (2004)]

B

B Horizon (B层, 沉淀层) (B céng, chén diàn céng)

The mineral horizon below the A horizon. The B horizon is in part a layer of transition from the overlying A to the underlying C horizon. The B horizon also has distinctive characteristics, such as (1) accumulation of clay, sesquioxides, humus, or a combination of these; (2) prismatic or block structure; (3) redder or browner colors than those in the A horizon; or (4) a combination of these. [Urban Soil Primer, p. 72 (2005)]

Backfill (回填材料) (huí tián cái liào)

Material used to refill excavated portions of a repository after waste has been emplaced. [IAEA Safety Glossary, p. 29 (2007)]

Bali Road Map (巴厘岛路线图) (bā lí dǎo lù xiàn tú)

A number of forward-looking decisions that represent various tracks that are essential to reaching a secure climate future, adopted at the 2007 Bali Climate Change Conference. [Bali Road Map (2012)]

Banned Chemical (禁用化学品) (jìn yòng huà xué pǐn)

A chemical all uses of which within one or more categories have been prohibited by final regulatory action, in order to protect human health or the environment. It includes a chemical that has been refused approval for first-time use or has been withdrawn by industry either from the domestic market or from further consideration in the domestic approval process and where there is clear evidence that such action has been taken in order to protect human health or the environment. [Rotterdam Convention, Art. 2 (1998)]

Basel Convention on the Control Of Transboundary Movements of Hazardous Wastes and their Disposal (关于控制危险废物越境转移及其处置的巴塞尔公约) (guān yú kòng zhìwēi xiǎn fèi wù yuè jìng zhuǎn yí jí qí chù zhì de bā sāi ěr gōng yuē)

A treaty adopted on March 22, 1989 [and entered into force on May 5, 1992]. The overarching objective of the Basel Convention is to protect human health and the environment against the adverse effects of hazardous wastes. Its scope of application covers a wide range of wastes defined as "hazardous wastes" based on their origin and/or composition and their characteristics, as well as two types of wastes defined as "other wastes" – household waste and incinerator ash. [Basel Convention (1989)]

Basel Protocol on Liability and Compensation (巴塞尔公约责任与赔偿议定书) (bā sāi ěr gōng yuē zé rèn yǔ péi cháng yì dìng shū)

A protocol to the Basel Convention adopted on December 10, 1999, addressing liability and compensation for damage resulting from the transboundary movement of hazardous wastes. [Basel Protocol (1999)]

Basel, Switzerland, Sandoz (Sandoz) Chemical Pollution of the Rhine Accident (瑞士巴塞尔桑多斯 (Sandoz) 化学公司莱茵河污染事故) (ruì shì bā sāi ěr sāng duō sī (Sandoz) huà xué gōng sī lái yīn hé wū rǎn shì gù)

A November 1, 1986, toxic agrochemical spill near Basel, Switzerland at the Sandoz storehouse causing the Rhine to turn red.

Basin (流域) (liú yù)

A natural depression on the Earth's surface, typically containing water.

Bay (海湾) (hǎi wān)

A broad inlet of a body of water where the land curves inward.

Becquerel (BQ) (贝可) (bèi kě)

The SI unit of activity, equal to one transformation per second. Supersedes the non-SI unit curie (Ci). 1 Bq = 27 pCi (2.7×10^{-11} Ci) approximately. 1 Ci = 3.7×10^{10} Bq. [IAEA Safety Glossary, p. 30 (2007)]

Bedrock (基岩) (jī yán)

The solid rock that underlies the soil and other unconsolidated material or that is exposed at the surface. [Urban Soil Primer, p. 71 (2005)]

Beef Hormone Case (United States and Canada v. European Union, 1998) (WT/DS26, WT/DS48) (牛肉荷尔蒙案 (美国、加拿大诉欧盟, 1988) (WT/DS26、WT/DS48)) (niú ròu hé ěr méng àn (měi guó, jiā ná dà sù ōu méng, 1998) (WT/DS26、WT/DS48))

A GATT panel decision striking down the EC's ban on the import of meat and meat products treated with specific growth hormones. The WTO Appellate Body upheld the WTO panel's ruling in favor of the United States and Canada.

Belgian Maas Valley Smog (比利时马斯河谷烟雾事件) (bǐ lì shí mǎ sī hégǔ yān wù shì jiàn)

In December 1930, a thermal inversion trapped smog in the Maas Valley for several days and killed more than 60 people within a week.

Benefit Sharing (惠益共享) (huì yì gòng xiǎng)

The equitable sharing of benefits, on agreed terms, arising from the use of biological and/or genetic material, with the providers of the material. [UNU IAS Pocket Guide, p. 10 (2007)]

Benzene (苯) (běn)

A cancer-causing hydrocarbon (C_6H_6) derived from petroleum. Benzene is a component of gasoline. Benzene emissions occur in exhaust as a byproduct of fuel combustion and also occur when gasoline evaporates. [US EPA Glossary of Mobile Source Emissions Terms (2012)]

Berlin Rules on Water Resources
(关于水资源的柏林规则 (简称柏林规则))
(guān yú shuǐ zī yuán de bó lín guī zé (jiǎn chēng bó lín guī zé))

A summary of the modern customary international law applicable to fresh water resources approved by the International Law Association. [Int'l Law Association (2004)]

Best Available Technology Economically Achievable (BAT)
(经济可行最佳可得技术) (jīng jì kě xíng zuì jiā kě dé jì shù)

Technology-based standard established by the Clean Water Act (CWA) as the most appropriate means available on a national basis for controlling the direct discharge of toxic and nonconventional pollutants to navigable waters. BAT effluent limitations guidelines, in general, represent the best existing performance of treatment technologies that are economically achievable within an industrial point source category or subcategory. [US EPA Glossary of NPDES Terms (2004)]

Best Conventional Pollutant Control Technology (BCT)
(最佳常规污染物控制技术) (zuì jā cháng guī wū rǎn wù kòng zhì jì shù)

Technology-based standard for the discharge from existing industrial point sources of conventional pollutants including BOD [Biochemical Oxygen Demand], TSS [Total Suspended Solids], fecal coliform, pH, oil and grease. The BCT is established in light of a two-part "cost reasonableness" test which compares the cost for an industry to reduce its pollutant discharge with the cost to a POTW [Publicly Owned Treatment Works] for similar levels of reduction of a pollutant loading. The second test examines the cost-effectiveness of additional industrial treatment beyond BPT. EPA must find limits which are reasonable under both tests before establishing them as BCT. [US EPA Glossary of NPDES Terms (2004)]

Best Management Practices (BMPs) (最佳管理惯例) (zuì jiā guǎn lǐ guàn lì)

Schedules of activities, prohibitions of practices, maintenance procedures, and other management practices to prevent or reduce the discharge of pollutants to waters of the United States. BMPs also include treatment requirements, operating procedures, and practice to control plant site runoff, spillage or leaks, sludge or waste disposal, or drainage from raw material storage. [US EPA Glossary of NPDES Terms (2004)]

Best Practicable Control Technology Currently Available (BPT) (当前可得最佳实用控制技术) (dāng qián kě dězuì jiā shí yòng kòng zhì jì shù)

The first level of technology-based standards established by the CWA to control pollutants discharged to waters of the . . . [US] BPT effluent limitations guidelines are generally based on the average of the best existing performance by plants within an industrial category or subcategory. [US EPA Glossary of NPDES Terms (2004)]

Best Professional Judgment (BPJ) (最佳职业判断) (zuì jā zhí yè pàn duàn)

The method used by permit writers to develop technology-based NPDES permit conditions on a case-by-case basis using all reasonably available and relevant data. [US EPA Glossary of NPDES Terms (2004)]

Bhopal, India Contamination (印度博帕尔污染案) (yìn dù bó pà ěr wū rǎn àn)/India's Bhopal Pesticide Spill (印度博帕尔农药泄漏事件) (yìn dù bó pà ěr nóng yào xiè lòu shì jiàn)

A December 2, 1984, industrial disaster caused by a gas leak at a pesticide factory in Bhopal, India. A toxic gas plume from a Union Carbide pesticide plant drifted over Bhopal, killing 2000 people almost instantly with thousands more dying later. The gas leak led to the deaths of more than 5000 people and continuing illness of 500 000 people.

Bigeye Tuna (大眼金枪鱼) (dà yǎn jīn qiāng yú)

Thunnus obesus. [Law of the Sea Convention, Annex 1, No. 3 (1982)]

Bioassay (生物测定) (shēng wù cè dìng)

(1) Any procedure used to determine the nature, activity, location, or retention of radionuclides in the body by direct (in vivo) measurement or by in vitro analysis of material excreted or otherwise removed from the body. [IAEA Safety Glossary, p. 30 (2007)] (2) A test used to evaluate the relative potency of a chemical or a mixture of chemicals by comparing its effect on a living organism with the effect of a standard preparation on the same type of organism. [US EPA Glossary of NPDES Terms (2004)]

Biochemical Oxygen Demand (BOD) (生化需氧量) (shēng huà xū yǎng liàng)

A measurement of the amount of oxygen utilized by the decomposition of organic material, over a specified time period (usually 5 days) in a waste-water sample; it is used as a measurement of the readily decomposable organic content of a wastewater. [US EPA Glossary of NPDES Terms (2004)]

Biodiversity Loss (生物多样性丧失) (shēng wù duō yàng xìng sàng shī)

A decrease in the variation of life forms on the planet. See **Biological Diversity**.

Bioenergy (生物能源) (shēng wù néng yuán)

Renewable energy generated from the conversion of biomass to energy.

Biogeochemical Cycle (生物地质化学循环) (shēng wù dì zhì huà xué xún huán)

Movements through the Earth system of key chemical constituents essential to life, such as carbon, nitrogen, oxygen, and phosphorus. [US EPA Glossary of Climate Change Terms (2011)]

Biological Contaminants (生物污染物) (shēng wù wū rǎn wù)

Agents derived from, or that are, living organisms (e.g., viruses, bacteria, fungi, and mammal and bird antigens) that can be inhaled and can cause many types of health effects including allergic reactions, respiratory disorders, hypersensitivity diseases, and infectious diseases. Also referred to as "microbiologicals" or "microbials." [US EPA Glossary of Indoor Air Quality Terms (2011)]

Biological Diversity (生物多样性) (shēng wù duō yàng xìng)

(1) The variability among living organisms from all sources, including, inter alia, terrestrial, marine, and other aquatic ecosystems and the ecological conplexes of which they are part: this includes diversity within species, between species, and of ecosystems. [Convention on Biological Diversity, Art. 2 (1992)] (2) The variability among living organisms from all sources including terrestrial, marine and other aquatic ecosystems. The diversity includes variability within species (genetic diversity), between species (species diversity), and ecosystems (ecosystem diversity). [UNU IAS Pocket Guide, p. 11 (2007)]

Biological Resources (生物资源) (shēng wù zī yuán)

Includes genetic resources, organisms or parts thereof, populations, or any other biotic component of ecosystems with actual or potential use or value for humanity. [Convention on Biological Diversity, Art. 2 (1992)]

Biomass (生物量) (shēng wù liàng)

Total dry weight of all living organisms that can be supported at each . . . [trophic] level in a food chain. Also, materials that are biological in origin, including organic material (both living and dead) from above and below ground, for example, trees, crops, grasses, tree litter, roots, and animals and animal waste. [US EPA Glossary of Climate Change Terms (2011)]

Biopiracy (生物剽窃) (shēng wù piáo qiè)

The appropriation of biological resources without prior informed consent of owners or local people or government. [UNU IAS Pocket Guide, p. 11 (2007)]

Bioprospecting (生物勘探) (shēng wù kān tàn)

The collection, research and use of biological and/or genetic material for purposes of applying the knowledge derived therefrom for scientific and/ or commercial purposes. Bioprospecting entails the search for economically valuable genetic and biochemical resources from nature. [UNU IAS Pocket Guide, p. 10 (2007)]

Biosolids (有机污泥) (yǒu jī wū ní)

Sewage sludge that is used or disposed through land application, surface disposal, incineration, or disposal in a municipal solid waste landfill. Sewage sludge is defined as solid, semi-solid, or liquid untreated residue generated during the treatment of domestic sewage in a treatment facility. [US EPA Glossary of NPDES Terms (2004)]

Biosphere (生物圈) (shēng wù quān)

(1) That part of the environment normally inhabited by living organisms. In practice, the biosphere is not usually defined with great precision, but is generally taken to include the atmosphere and the Earth's surface, including the soil and surface water bodies, seas and oceans and their sediments. There is no generally accepted definition of the depth below the surface at which soil or sediment ceases to be part of the biosphere, but this might typically be taken to be the depth affected by basic human actions, in particular farming. In waste safety, in particular, the biosphere is normally distinguished from the geosphere. [IAEA Safety Glossary, p. 30 (2007)] (2)

The part of the Earth system comprising all ecosystems and living organisms, in the atmosphere, on land (terrestrial biosphere) or in the oceans (marine biosphere), including derived dead organic matter, such as litter, soil organic matter, and oceanic detritus. [US EPA Glossary of Climate Change Terms (2011)]

Biotechnology (生物技术) (shēng wù jì shù)

Any technological application that uses biological systems, living organisms, or derivatives thereof, to make or modify products or processes for specific use. [Convention on Biological Diversity, Art. 2 (1992)]

Biotransformation (生物转化) (shēng wù zhuǎn huà)

The alteration of a substance into another using a biological catalyst.

Black Carbon (炭黑) (tàn hēi)

Operationally defined species [of particulate air pollution] based on measurement of light absorption and chemical reactivity and/or thermal stability; consists of soot, charcoal, and/or possible light-absorbing refractory organic matter. (Source Charlson and Heintzenberg, 1995, p. 401.) [US EPA Glossary of Climate Change Terms (2011)]

Blackfin Tuna (黑鳍金枪鱼) (hēi qí jīn qiāng yú)

Thunnus atlanticus. [Law of the Sea Convention, Annex 1, No. 6 (1982)]

Bluefin Tuna (蓝鳍金枪鱼) (lán qí jīn qiāng yú)

Thunnus thynnus. [Law of the Sea Convention, Annex 1, No. 2 (1982)]

Borehole (钻孔) (zuān kǒng)

Any exploratory hole drilled into the Earth or ice to gather geophysical data. Climate researchers often take ice core samples, a type of borehole, to predict atmospheric composition in earlier years. [US EPA Glossary of Climate Change Terms (2011)]

Botany (植物学) (zhí wù xué)

The scientific study of plants.

Breadth of the Exclusive Economic Zone (专属经济区宽度) (zhuān shǔ jīng jì qū kuān dù)

The exclusive economic zone shall not extend beyond 200 nautical miles from the baselines from which the breadth of the territorial sea is measured. [Law of the Sea Convention, Art. 57 (1982)]

Breadth of the Territorial Sea (领海宽度) (lǐng hǎi kuān dù)

Every State has the right to establish the breadth of its territorial sea up to a limit not exceeding 12 nautical miles, measured from baselines determined in accordance with this Convention. [Law of the Sea Convention, Art. 3 (1982)]

Breathing Zone (呼吸区) (hū xī qū)

The breathing zone is within a ten inch radius of the worker's nose and mouth. [OSHA Glossary 2012]

Bubble Policy (泡泡政策) (pào pào zhèng cè)

EPA policy that allows a plant complex with several facilities to decrease pollution from some facilities while increasing it from others, so long as total results are equal to or better than previous limits. Facilities where this is done are treated as if they exist in a bubble in which total emissions are averaged out. [GEMET Thesaurus (2012)]

Buffer (缓冲物质) (huǎn chōng wù zhì)

(1) Any substance placed around a waste package in a repository to serve as a barrier to restrict the access of groundwater to the waste package and to reduce by sorption and precipitation the rate of eventual migration of radionuclides from the waste. The above definition is clearly specific to waste safety. The term buffer (e.g., in buffer solution) is also used, in its normal scientific sense (and therefore normally without specific definition), in a variety of contexts. [IAEA Safety Glossary, p. 30 (2007)] (2) A buffer is a designated area of land that is controlled by local regulations to protect an adjacent area from the impacts of development. [Nolon, Well Grounded, p. 446 (ELI 2001)]

Building Area (建筑面积) (jiàn zhù miàn jī)

The total square footage of a parcel of land that is allowed by the regulations to be covered by buildings and other physical improvements. [Nolon, Well Grounded, p. 446 (ELI 2001)]

Building Code (建筑规范) (jiàn zhù guī fàn)

The Uniform Fire Prevention and Building Code, as modified by local amendments. This code governs the construction details of buildings and other structures in the interests of the safety of the occupants and the public. A local building inspector may not issue a building permit unless the applicant's construction drawings comply with the provisions of the building code. [Nolon, Well Grounded, p. 446 (ELI 2001)]

Building Energy Conservation/Building Energy Efficiency (建筑节能) (jiàn zhù jié néng)

The reduction of energy consumption by a building through the use of more energy efficient design and technologies.

Building Envelope (围护结构) (wéi hù jié gòu)

Elements of the building, including all external building materials, windows, and walls, that enclose the internal space. [US EPA Glossary of Indoor Air Quality Terms (2011)]

Building Height (建筑高度) (jiàn zhù gāo dù)

The vertical distance from the average elevation of the proposed finished grade along the wall of a building or structure to the highest point of the roof, for flat roofs, or to the mean height between eaves and ridge, for gable, hip, and gambrel roofs. [Nolon, Well Grounded, p. 446 (ELI 2001)]

Building Inspector (建筑检查员) (jiàn zhù jiǎn chá yuán)

The local administrative official charged with the responsibility of administering and enforcing the provisions of the building code. In some communities the building inspector may also be the zoning enforcement officer. [Nolon, Well Grounded, p. 446 (ELI 2001)]

Building Permit (建筑执照) (jiàn zhù zhí zhào)

A building permit must be issued by a municipal agency or officer before activities such as construction, alteration, or expansion of buildings or improvements on the land may legally commence. [Nolon, Well Grounded, p. 446 (ELI 2001)]

Building-related Illness (BRI) (大楼并发症症) (dà lóu bìng fā zhèng)

Diagnosable illness whose symptoms can be identified and whose cause can be directly attributed to airborne building pollutants (e.g., Legionnaire's disease, hypersensitivity pneumonitis). Also: . . . [a] discrete, identifiable disease or illness that can be traced to a specific pollutant or source within a building. [US EPA Glossary of Indoor Air Quality Terms (2011)]

Bulk Regulations (容积规章) (róng jī guī zhāng)

The controls in a zoning district governing the size, location, and dimensions of buildings and improvements on a parcel of land. [Nolon, Well Grounded, p. 446 (ELI 2001)]

Burnable Absorber (可燃吸收剂) (kě rán xī shōu jì)

Neutron absorbing material, used to control reactivity, with particular capability of being depleted by neutron absorption. [IAEA Safety Glossary, p. 31 (2007)]

Bypass (分流) (fēn líu)

The intentional diversion of wastestreams from any portion of a treatment (or pretreatment) facility. [US EPA Glossary of NPDES Terms (2004)]

C

C Horizon (母质层) (mǔ zhì céng)

C horizons are below B horizons and are commonly referred to as the substratum. They are made up mainly of partially weathered or disintegrated parent material, but soft bedrock can also occur. Because C horizons are deeper in the profile, the effects of the soil-forming factors are less pronounced than the effects in the overlying A and B horizons. [Urban Soil Primer, p. 13 (2005)]

Calculated Levels (计算数量) (jì suàn shù liàng)

Calculated levels of production, imports, exports and consumption means levels determined in accordance with Article 3 [Calculation of control levels]. [Montreal Protocol, Art. 1, No. 7 (1987)]

Calvert Cliffs' Coordinating Committee v. U.S. Atomic Energy Commission (卡尔弗特·克利夫协调委员会诉美国原子能委员会案) (kǎ ěr fú tè ·kè lì fū xié tiào wěi yuán huì sù měi guó yuán zǐ néng wěi yuán huì àn)

A 1971 decision by the US Court of Appeals, District of Columbia Circuit, holding that courts have power to require agencies to comply with procedural directions of National Environmental Policy Act of 1969 and that the Commission's rules did not comply with the Act when precluding review consideration of nonradiological environmental issues unless specifically raised, prohibiting raising such issues in certain pending proceedings or when issues have been passed on by other agencies, and precluding consideration between grant of construction permit and consideration of grant of operating license. [Calvert Cliffs' 1971]

Capital Budget (资本预算) (zī běn yù suàn)

The municipal budget that provides for the construction of capital projects in the community. [Nolon, Well Grounded, p. 447 (ELI 2001)]

Capital Project (基本工程项目) (jī běn gōng chéng xiàng mù)

Construction projects including public buildings, roads, street improvements, lighting, parks, and their improvement or rehabilitation paid for under the community's capital budget. [Nolon, Well Grounded, p. 447 (ELI 2001)]

Car Tax Case (E.C. v. US) (DS31/R, 1994)
(汽车税案 (欧共体诉美国) (DS31/R, 1994)) (qì chē shuì àn (ōu gòng tǐ sù měi guó) (DS31/R, 1994))

A 1994 case before the General Agreement on Tariffs and Trade Council regarding three United States car taxes: the luxury tax, corporate average fuel economy (CAFE) standard tax, and gas guzzler tax. [United States – Taxes on Automobiles (1994)]

Carbon Cycle (碳循环) (tàn xún huán)

All parts (reservoirs) and fluxes of carbon. The cycle is usually thought of as four main reservoirs of carbon interconnected by pathways of exchange. The reservoirs are the atmosphere, terrestrial biosphere (usually includes freshwater systems), oceans, and sediments (includes fossil fuels). The annual movements of carbon, the carbon exchanges between reservoirs, occur because of various chemical, physical, geological, and biological processes. The ocean contains the largest pool of carbon near the surface of the Earth, but most of that pool is not involved with rapid exchange with the atmosphere. [US EPA Glossary of Climate Change Terms (2011)]

Carbon Dioxide (二氧化碳) (èr yǎng huà tàn)

(1) The main greenhouse gas caused by human activities; it also originates from natural sources, like volcanic activity. [UNEP Guide to Climate Neutrality, p. 194 (2008)] (2) A naturally occurring gas, and also a by-product of burning fossil fuels and biomass, as well as land-use changes and other industrial processes. It is the principal anthropogenic greenhouse gas that affects the Earth's radiative balance. It is the reference gas against which other greenhouse gases are measured and therefore has a Global Warming Potential of 1. [US EPA Glossary of Climate Change Terms (2011)] See **Climate Change** and **Global Warming**.

Carbon Dioxide Equivalent (二氧化碳当量) (èr yǎng huà tàn dāng liàng)

A metric measure used to compare the emissions from various greenhouse gases based upon their global warming potential (GWP). Carbon dioxide equivalents are commonly expressed as "million metric tons of carbon dioxide equivalents (MMTCO$_2$Eq)." The carbon dioxide equivalent for a gas is derived by multiplying the tons of the gas by the associated GWP. The use of carbon equivalents (MMTCE) is declining. [US EPA Glossary of Climate Change Terms (2011)]

Carbon Dioxide Fertilization (二氧化碳施肥) (èr yǎng huà tàn shī féi)

The enhancement of the growth of plants as a result of increased atmospheric CO$_2$ concentration. Depending on their mechanism of photosynthesis, certain types of plants are more sensitive to changes in atmospheric CO$_2$ concentration. [US EPA Glossary of Climate Change Terms (2011)]

Carbon Intensity (碳强度) (tàn qiáng dù)

The amount of carbon by weight emitted per unit of energy consumed. A common measure of carbon intensity is weight of carbon per British thermal unit (Btu) of energy. When there is only one fossil fuel under consideration, the carbon intensity and the emissions coefficient are identical. When there are several fuels, carbon intensity is based on their combined emissions coefficients weighted by their energy consumption levels. [US EPA Glossary of Climate Change Terms (2011)]

Carbon Monoxide (CO) (一氧化碳) (yī yǎng huà tàn)

A colorless, odorless gas that forms when carbon in fuel is not burned completely. Carbon monoxide is a component of exhaust from motor vehicles and engines. Carbon monoxide emissions increase when conditions are poor for combustion; thus, the highest carbon monoxide levels tend to occur when the weather is very cold or at high elevations where there is less oxygen in the air to burn the fuel. [US EPA Glossary of Climate Change Terms (2011)]

Carbon Sequestration (碳封存) (tàn fēng cún)

The uptake and storage of carbon. Trees and plants, for example, absorb carbon dioxide, release the oxygen and store the carbon. Fossil fuels were at one time biomass and continue to store the carbon until burned. [US EPA Glossary of Climate Change Terms (2011)]

Carbon Sequestration and Storage (碳吸收与储存) (tàn xī shōu yǔ chǔ cún)

An experimental technology designed to remove carbon dioxide from emissions such as power stations: the gas is then liquified and pumped into rock formations underground or beneath the sea bed. Proponents believe it has great potential for tackling climate change but CCS is not yet available at a commercial stage. [UNEP Guide to Climate Neutrality, p. 194 (2008)]

Carbon Sink (碳汇) (tàn huì)

A natural feature – a forest, for example, or a peat bog – which absorbs CO_2. [UNEP Guide to Climate Neutrality, p. 194 (2008)]

Carbon Trading (碳交易) (tàn jiāo yì)

(1) A market established by law and regulation whereby limitations are established to cap emissions of carbon dioxide from the combustion of fossil fuels, and those who reduce their emissions more than is required may trade or sell their extra reductions to others who have failed to meet their own emission reduction requirements and who can satisfy their requirements by acquiring the extra amounts to satisfy their reduction obligations. (2) A market based mechanism for helping mitigate the increase of CO_2 in the atmosphere. . . . [The goal of these markets is to] bring buyers and sellers of carbon credits together with standardized rules of trade. [What is Carbon Trading?]

Carrier (承运人) (chéng yùn rén)

(1) Any person who carries out the transport of hazardous wastes or other wastes. [Basel Convention, Art. 2, No. 17 (1989)] (2) Any person, organization or government undertaking the carriage of radioactive material by any means of transport. The term includes both carriers for hire or reward (known as common or contract carriers in some countries) and carriers on own account (known as private carriers in some countries). [IAEA Safety Glossary, p. 32 (2007)]

Cartagena Protocol on Biosafety (卡塔赫纳生物安全议定书) (kǎ tǎ hè nà shēng wù ān quán yì dìng shū)

An international agreement which aims to ensure the safe handling, transport and use of living modified organisms (LMOs) resulting from modern biotechnology that may have adverse effects on biological diversity, taking also into account risks to human health. The agreement was adopted on January 29, 2000, and entered into force on September 11, 2003. [Cartagena Protocol]

Catalytic Converter (催化转化器) (cuī huà zhuǎn huà qì)

An anti-pollution device located between a vehicle's engine and tailpipe. Catalytic converters work by facilitating chemical reactions that convert exhaust pollutants such as carbon monoxide and nitrogen oxides to normal atmospheric gases such as nitrogen, carbon dioxide, and water. [US EPA Glossary of Climate Change Terms (2011)]

Categorical Industrial User (CIU) (工业类用户) (gōng yè lèi yòng hù)

An industrial user subject to national categorical pretreatment standards. [US EPA Glossary of NPDES Terms (2004)]

Categorical Pretreatment Standards (预处理类标准) (yù chù lǐ lèi biāo zhǔn)

Limitations on pollutant discharges to publicly owned treatment works promulgated by EPA in accordance with Section 307 of the Clean Water Act that apply to specified process wastewaters of particular industrial categories. [US EPA Glossary of NPDES Terms (2004)]

Catena (土壤链) (tǔ rǎng liàn)

A sequence, or "chain," of soils on a landscape that formed in similar kinds of parent material but have different characteristics as a result of differences in relief and drainage. [Urban Soil Primer, p. 71 (2005)]

Cation (阳离子) (yáng lí zǐ)

An ion carrying a positive charge of electricity. The common soil cations are calcium, potassium, magnesium, sodium, and hydrogen. [Urban Soil Primer, p. 71 (2005)]

Cation-exchange Capacity (阳离子交换能力) (yáng lí zǐ jiāo huàn néng lì)

The total amount of exchangeable cations that can be held by the soil, expressed in terms of milliequivalents per 100 grams of soil at neutrality (pH 7.0) or at some other stated pH value. [Urban Soil Primer, p. 71 (2005)]

Causal Relation (因果关系) (yīn guǒ guān xì)

The link between one action that produces change or impact in another action, or the link between one action and a result.

Ceiling Plenum Space (天花板静压空间) (tiān huā bǎn jìng yā kōng jiān)

Space below the flooring and above the suspended ceiling that accommodates the mechanical and electrical equipment and that is used as part of the air distribution system. The space is kept under negative pressure. [US EPA Glossary of Indoor Air Quality Terms (2011)]

Cellular Facility (移动通讯设施) (yí dòng tōng xùn shè shī)

An individual cell of a cellular transmission system that includes a base station, antennae, and associated electronic equipment that sends to and receives signals from mobile phones. [Nolon, Well Grounded, p. 447 (ELI 2001)]

Central Air Handling Unit (Central AHU) (中央空气处理机组) (zhōng yāng kōng qì chù lǐ jī zǔ)

This is the same as an Air Handling Unit, but serves more than one area. [US EPA Glossary of Indoor Air Quality Terms (2011)]

Centralized Wastewater Treatment System (污水集中处理系统) (wū shuǐ jí zhōng chǔ lǐ xì tǒng)

A managed system consisting of collection sewers and a single treatment plant used to collect and treat wastewater from an entire service area. Traditionally, such a system has been called a publicly owned treatment works (POTW) as defined at 40 CFR 122.2. [US EPA Glossary of NPDES Terms (2004)]

Certification (认证) (rèn zhèng)

Procedure by which a third party gives written assurance that a product, process, or service conforms to specified requirements. [Von Zharen, ISO 14000, p. 194 (1996)]

Certified (经认证的) (jīng rèn zhèng de)

The Environmental Management System (EMS) of a company, location, or plant is certified for compliance to ISO 14001 after it has demonstrated such compliance through the audit process. When used to indicate EMS certification, it means the same thing as registration. [Von Zharen, ISO 14000, p. 194 (1996)]

Certifying Official (认证官员) (rèn zhèng guān yuán)

All applications, including Notices of Intent (NOIs), must be signed as follows.

1. For a corporation: By a responsible corporate officer. For the purpose of this Part, a responsible corporate officer means: (i) a president, secretary, treasurer, or vice-president of the corporation in charge of a principal business function, or any other person who performs similar policy- or decision-making functions for the corporation, or (ii) the manager of one or more manufacturing, production, or operating facilities, provided, the manager is authorized to make management decisions which govern the operation of the regulated facility including having the explicit or implicit duty of making major capital investment recommendations, and initiating and directing other comprehensive measures to assure long term environmental compliance with environmental laws and regulations; the manager can ensure that the necessary systems are established or actions taken to gather complete and accurate information for permit application requirements; and where authority to sign documents has been assigned or delegated to the manager in accordance with corporate procedures.
2. For a partnership or sole proprietorship: By a general partner or the proprietor, respectively; or
3. For a municipality, state, federal, or other public agency: By either a principal executive officer or ranking elected official. For purposes of this Part, a principal executive officer of a federal agency includes (i) the chief executive officer of the agency, or (ii) a senior executive officer having responsibility for the overall operations of a principal geographic unit of the agency (e.g., Regional Administrator of EPA). [US EPA Glossary of NPDES Terms (2004)]

CFM (立方英尺/分钟) (lì fāng yīng chǐ/fēn zhōng)

Cubic feet per minute. The amount of air, in cubic feet, that flows through a given space in one minute; 1 CFM equals approximately 2 liters per second (l/s). [US EPA Glossary of Indoor Air Quality Terms (2011)]

Chemical (化学品) (huà xué pǐn)

A substance whether by itself or in a mixture or preparation and whether manufactured or obtained from nature, but does not include any living organism. It consists of the following categories: pesticide (including severely hazardous pesticide formulations) and industrial. [Rotterdam Convention, Art. 2 (1998)]

Chemical Oxygen Demand (COD) (化学需氧量) (huà xué xū yǎng liàng)

A measure of the oxygen-consuming capacity of inorganic and organic matter present in wastewater. COD is expressed as the amount of oxygen consumed in mg/l. Results do not necessarily correlate to the biochemical oxygen demand (BOD) because the chemical oxidant may react with substances that bacteria do not stabilize. [US EPA Glossary of NPDES Terms (2004)]

Chemical Review Committee (化学品审查委员会) (huà xué pǐn shěn chá wěi yuán huì)

The subsidiary body referred to in paragraph 6 of Article 18 of the Rotterdam Convention. [Rotterdam Convention, Art. 2 (1998)]

Chemical Sensitization (化学过敏) (huà xué guò mǐn)

Evidence suggests that some people may develop health problems characterized by effects such as dizziness, eye and throat irritation, chest tightness, and nasal congestion that appear whenever they are exposed to certain chemicals. People may react to even trace amounts of chemicals to which they have become "sensitized." [US EPA Glossary of Indoor Air Quality Terms (2011)]

Chernobyl Nuclear Accident (切尔诺贝利核电站泄漏事故) (qiè ěr nuò bèi lì hé diàn zhàn xiè lù shì gù)

A 1986 nuclear accident on the border of Ukraine with Belarus, resulting in dozens of deaths and hundreds of cases of acute radiation syndrome (ARS).

Chevron U.S.A., Inc. v. Natural Resources Defense Council (美国雪弗龙公司诉自然资源保卫委员会案 (měi guó xuě fú lóng gōng sī sù zì rán zī yuán bǎo wèi wěi yuán huì àn)

A 1984 US Supreme Court case that set forth the test for determining the scope of judicial deference to an administrative agency's interpretation of statutes that the agency administers.

Chlorofluorocarbons (氯氟化碳) (lù fú huà tàn)

Greenhouse gases covered under the 1987 Montreal Protocol and used for refrigeration, air conditioning, packaging, insulation, solvents, or aerosol propellants. Since they are not destroyed in the lower atmosphere, CFCs drift into the upper atmosphere where, given suitable conditions, they break down ozone. These gases are being replaced by other compounds, including hydrochlorofluorocarbons and hydrofluorocarbons, which are greenhouse gases covered under the Kyoto Protocol. [US EPA Glossary of Climate Change Terms (2011)]

Chronic Effect (慢性效应) (màn xìng xiào yìng)

(1) Adverse effect on a living being that develops slowly from a man-made or natural toxin or allergen (as compared to an *acute* effect, that develops rapidly). (2) A stimulus that lingers or continues for a relatively long period of time, often one-tenth of the life span or more. Chronic should be considered a relative term depending on the life span of an organism. The measurement of a chronic effect can be reduced growth, reduced reproduction, etc., in addition to lethality. [US EPA Glossary of NPDES Terms (2004)]

Circular Economy, And Waste Disposal Laws
(循环经济和废物处置法) (xún huán jīng jì hé fèi wù chǔ zhì fǎ)

Chinese laws providing that all materials and energy in production cycles are valuable and shall be used and reused, or recycled, and not allowed to become waste; any residue shall be stored safely and conserved for future recycling.

Civil Litigation (民事诉讼) (mín shì sù sòng)

An adjudication procedure in a judicial court in which two or more individuals, organizations, enterprises or agencies and authorities present their claims and defenses in order to obtain a decision that resolves their disputes in an orderly and peaceful manner.

Class V Injection Well (V类注水井) (V lèi zhù shuǐ jǐng)

A shallow well used to place a variety of fluids at shallow depths below the land surface, including a domestic onsite wastewater treatment system serving more than 20 people. USEPA permits these wells to inject wastes below the ground surface provided they meet certain requirements and do not endanger underground sources of drinking water. [US EPA Glossary of NPDES Terms (2004)]

Clay (粘土) (nián tǔ)

As a soil separate, the mineral soil particles less than 0.002 mm in diameter. As a soil textural class, soil material that is 40 percent or more clay, less than 45 percent sand, and less than 40 percent silt. [Urban Soil Primer, p. 71 (2005)]

Clean Air Act (CAA, United States)
(清洁空气法) (CAA, 美国) (qīng jié kōng qì fǎ (měi guó))

The legal authority for US federal programs regarding air pollution control. The basis of this authority is found in the 1990 Clean Air Act Amendments. [US EPA, Clean Air Act (2012)]

Clean Development Mechanism (CDM)
(清洁发展机制 (CDM)) (qīng jié fā zhǎn jī zhì (CDM))

(1) The techniques agreed under the terms of the United Nations Framework Convention on Climate Change by which States propose sustainable development projects that can be financed through funding that provides other States with a credit or off-set for the emission of greenhouse gases. (2) The *CDM* allows emission-reduction projects in developing countries to earn certified emission reduction (CER) credits, each equivalent to one ton of CO_2. These CERs can be traded and sold, and used by industrialized countries to a meet a part of their emission reduction targets under the Kyoto Protocol. The mechanism stimulates sustainable development and emission reductions, while giving industrialized countries some flexibility in how they meet their emission reduction limitation targets. The CDM is the main source of income for the UNFCCC Adaptation Fund, which was established to finance adaptation projects and . . . [programs] in developing country Parties to the Kyoto Protocol that are particularly vulnerable to the adverse effects of climate change. The Adaptation Fund is financed by a 2. . . [percent] levy on CERs issued by the CDM. [Clean Development Mechanism]

Clean Production Processes (清洁生产工艺) (qīng jié shēng chǎn gōng yì)

(1) The pollution control and other environmental management systems used in a manufacturing facility to ensure that the emissions of wastes are controlled and the use of energy and other resources is as efficient as possible. (2) Cleaner Production means the continuous application of an integrated preventive environmental strategy to processes and products to reduce risks to humans and the environment. For Production processes, cleaner production includes conserving raw materials and energy, eliminating toxic raw materials, and reducing the quantity and toxicity of all emissions and wastes before they leave a process. [Clean Production Action]

Clean Production Technologies (清洁生产技术) (qīng jié shēng chǎn jì shù)

The engineering and other physical elements of machinery and assembly line and other manufacturing systems that are used to control pollution and efficient use of energy and resources.

Clean Water Act (CWA, United States) (清洁水法 (CWA, 美国)) (qīng jié shuǐ fǎ (měi guó))

Short name for the 1972 amendments to the 1948 Federal Water Pollution Control Act. The CWA establishes the basic structure for regulating discharges of pollutants into the waters of the United States and regulating quality standards for surface waters. [Clean Water Act, 33 U.S.C. §1251 et seq. (1972)]

Client (委托人) (wěi tuō rén)

Organization commissioning the audit. The client may be the auditee, or any other organization which has the regulatory or contractual right to commission an audit. [Von Zharen, ISO 14000, p. 194 (1996)]

Climate (气候) (qì hòu)

Climate in a narrow sense is usually defined as the "average weather," or more rigorously, as the statistical description in terms of the mean and variability of relevant quantities over a period of time ranging from months to thousands of years. The classical period is . . .[three] decades, as defined by the World Meteorological Organization (WMO). These quantities are most often surface variables such as temperature, precipitation, and wind. Climate in a wider sense is the state, including a statistical description, of the climate system. [US EPA Glossary of Climate Change Terms (2011)]

Climate Change (气候变化) (qì hòu biàn huà)

(1) A change of climate which is attributed directly or indirectly to human activity that alters the composition of the global atmosphere and which is in addtition to natural climate variability observed over comparable time periods. [United Nations Framework Convention on Climate Change, Art. 1, No. 2 (1992)] (2) Climate change refers to any significant change in measures of climate (such as temperature, precipitation, or wind) lasting for an extended period (decades or longer). Climate change may result from natural factors, such as changes in the sun's intensity or slow changes in the Earth's orbit around the sun; natural processes within the climate system (e.g., changes in ocean circulation); human activities that change

the atmosphere's composition (e.g., through burning fossil fuels) and the land surface (e.g., deforestation, reforestation, urbanization, desertification, etc.). [US EPA Glossary of Climate Change Terms (2011)]

Climate Feedback (气候反馈) (qì hòu fǎn kuì)

An interaction mechanism between processes in the climate system is called a climate feedback, when the result of an initial process triggers changes in a second process that in turn influences the initial one. A positive feedback intensifies the original process, and a negative feedback reduces it. [US EPA Glossary of Climate Change Terms (2011)]

Climate Lag (气候滞后) (qì hòu zhì hòu)

The delay that occurs in climate change as a result of some factor that changes only very slowly. For example, the effects of releasing more carbon dioxide into the atmosphere may not be known for some time because a large fraction is dissolved in the ocean and only released to the atmosphere many years later.

Climate Model (气候模型) (qì hòu mó xíng)

A quantitative way of representing the interactions of the atmosphere, oceans, land surface, and ice. Models can range from relatively simple to quite comprehensive. [US EPA Glossary of Climate Change Terms (2011)]

Climate Sensitivity (气候敏感性) (qì hòu mǐn gǎn xìng)

In IPCC Reports, equilibrium climate sensitivity refers to the equilibrium change in global mean surface temperature following a doubling of the atmospheric (equivalent) CO_2 concentration. More generally, equilibrium climate sensitivity refers to the equilibrium change in surface air temperature following a unit change in radiative forcing (degrees Celsius, per watts per square meter, $°C/Wm^{-2}$). In practice, the evaluation of the equilibrium climate sensitivity requires very long simulations with Coupled General Circulation Models (Climate model). The effective climate sensitivity is a related measure that circumvents this requirement. It is evaluated from model output for evolving non-equilibrium conditions. It is a measure of the strengths of the feedbacks at a particular time and may vary with forcing history and climate state. [US EPA Glossary of Climate Change Terms (2011)]

Climate System (气候系统) (qì hòu xì tǒng)

The totality of the atmosphere, hydrosphere, biosphere and geosphere and their interactions. [United Nations Framework Convention on Climate Change, Art. 1, No. 3 (1992)]

Climate System (or Earth System) (气候系统 (或者地球系统)) (qì hòu xì tǒng (huò zhě dì qiú xì tǒng))

The five physical components (atmosphere, hydrosphere, cryosphere, lithosphere, and biosphere) that are responsible for the climate and its variations. [US EPA Glossary of Climate Change Terms (2011)]

CO_2 Equivalence (二氧化碳当量) (èr yǎng huà tàn dāng liàng)

A way of expressing the combined efficiency of all greenhouse gases: carbon dioxide (CO_2), methane (CH_4), nitrous oxide (N_2O), and the rarer trace greenhouse gases such as chlorofluorocarbons. Their potency varies according to their chemical makeup and the length of time they persist in the atmosphere. [UNEP Guide to Climate Neutrality, p. 194 (2008)]

Coal Mine Methane (煤矿瓦斯) (méi kuàng wǎ sī)

Coal mine methane is the subset of CBM that is released from the coal seams during the process of coal mining. [US EPA Glossary of Climate Change Terms (2011)] See **Coalbed Methane**.

Coalbed Methane (CBM) (煤层气) (méi céng qì)

Coalbed methane is methane contained in coal seams, and is often referred to as virgin coalbed methane, or coal seam gas. [US EPA Glossary of Climate Change Terms (2011)]

Co-benefit (共生效益) (gòng shēng xiào yì)

The benefits of policies that are implemented for various reasons at the same time – including climate change mitigation – acknowledging that most policies designed to address greenhouse gas mitigation also have other, often at least equally important, rationales (e.g., related to objectives of development, sustainability, and equity). The term co-impact is also used in a more generic sense to cover both the positive and negative side of the benefits. [US EPA Glossary of Climate Change Terms (2011)]

Code of Federal Regulations (CFR, United States) (美国联邦法规汇编) (měi guó lián bāng fǎ guī huì biān)

A codification of the final rules published daily in the Federal Register. Title 40 of the CFR contains the environmental regulations. [US EPA Glossary of NPDES Terms (2004)]

Combating Desertification (防治沙漠化) (fáng zhì shā mò huà)

Includes activities which are part of the integrated development of land in arid, semi-arid and dry sub-humid areas for sustainable development which are aimed at: (i) prevention and/or reduction of land degradation; (ii) rehabilitation of partly degraded land; and (iii) reclamation of desertified land. [UN Convention to Combat Desertification (1994)] See **Desertification**.

Combination Foundations (组合地基) (zǔ hé dì jī)

Buildings constructed with more than one foundation type; e.g., basement/crawlspace or basement/slab-on-grade. [US EPA Glossary of Indoor Air Quality Terms (2011)]

Combined Sewer Overflow (CSO) (合流下水道溢流) (hé liú xià shuǐ dào yì liú)

A discharge of untreated wastewater from a combined sewer system at a point prior to the headworks of a publicly owned treatment works. CSOs generally occur during wet weather (rainfall or snowmelt). During periods of wet weather, these systems become overloaded, bypass treatment works, and discharge directly to receiving waters. [US EPA Glossary of NPDES Terms (2004)]

Combined Sewer System (CSS) (合流制排水系统) (hé liú zhì pái shuǐ xì tǒng)

A wastewater collection system which conveys sanitary wastewaters (domestic, commercial and industrial wastewaters) and stormwater through a single pipe to a publicly owned treatment works for treatment prior to discharge to surface waters. [US EPA Glossary of NPDES Terms (2004)]

Combustion (燃烧) (rán shāo)

The process of burning. Motor vehicles and equipment typically burn fuel in an engine to create power. Gasoline and diesel fuels are mixtures of hydrocarbons, which are compounds that contain hydrogen and carbon atoms. In "perfect" combustion, oxygen in the air would combine with all the hydrogen in the fuel to form water and with all the carbon in the fuel to form carbon dioxide. Nitrogen in the air would remain unaffected. In reality, the combustion process is not "perfect," and engines emit several types of pollutants as combustion byproducts. [US EPA Glossary of Mobile Source Emissions Terms (2012)]

Commercial Use (商业用途) (shāng yè yòng tú)

Any use of biodiversity and/or genetic resources, their products or derivatives for monetary gains that includes selling in the market. [UNU IAS Pocket Guide, p. 12 (2007)]

Commissioning (试运转) (shì yùn zhuǎn)

Start-up of a building that includes testing and adjusting Heating, Ventilation, Air Conditioning System (HVAC), electrical, plumbing, and other systems to assure proper functioning and adherence to design criteria. Commissioning also includes the instruction of building representatives in the use of the building systems. [US EPA Glossary of Indoor Air Quality Terms (2011)]

Common but Differentiated Responsibilities
(共同但有区别的责任) (gòng tóng dàn yǒu qū bié de zé rèn)

A cornerstone of sustainable development, articulated in Principle 7 of the Rio Declaration. It provides, "In view of the different contributions to global environmental degradation, States have common but differentiated responsibilities. The developed countries acknowledge the responsibility that they bear in the international pursuit of sustainable development in view of the pressures their societies place on the global environment and of the technologies and financial resources they command." [Encyclopedia of Earth]

Common Heritage of Mankind (人类共同遗产)
(rén lèi gòng tóng yí chǎn)

Resources that exist across many regions and reflect values shared by all States and peoples, such as world cultural heritage sites, wetlands of international importance as habitat for migratory species, or areas such as the resources of the high seas beyond areas of national jurisdiction or of the Moon and resources in space beyond the Earth.

Common Human Concerns (人类共同关切事项)
(rén lèi gòng tóng guān qiè shì xiàng)

Values and interests shared by individuals and communities universally and sustained through the establishment of common or similar procedures that recognize these values and interests and seek to maintain them.

Common Interests (共同利益) (gòng tóng lì yì)

Values or dependencies that are shared by similarly situated individuals, enterprises, organizations, authorities and States or intergovernmental bodies.

Common Property of Mankind (人类共同财产) (rén lèi gòng tóng cái chǎn)

(1) The Law of the Sea Convention designates the high seas as the "common heritage of mankind," a principle sometimes referred to as the "common property of mankind." (2) The cultural or natural or genetic resources that States have agreed are held collectively or in trust as the property of all, and the use of which is allocated or permitted in accordance with internationally agreed procedures.

Compaction (压紧) (yā jǐn)

Creation of dense soil layers when the soil is subject to the heavy weight of machinery or foot traffic, especially during wet periods. [Urban Soil Primer, p. 71 (2005)]

Comparative Assertion (比较断言) (bǐ jiào duàn yán)

Environment claim regarding the superiority of one product versus a competing product. [Von Zharen, ISO 14000, p. 194 (1996)]

Compensation For Health Hazards (健康损害补偿) (jiàn kāng sǔn hài bǔ cháng)

Financial payments or services, such as annual medical health assessments and treatment as appropriate, provided to an individual for exposure to substances or activities that are or reasonably could be connected to adverse health consequences.

Competent Authority (主管当局) (zhǔ guǎn dāng jú)

One governmental authority designated by a Party to be responsible, within such geographical areas as the Party may think fit, for receiving the notification of a transboundary movement of hazardous wastes or other wastes, and any information related to it, and for responding to such a notification, as provided in Article 6 [Transboundary Movement Between Parties]. [Basel Convention, Art. 2, No. 6 (1989)]

Compliance (合格) (hé gé)

An affirmative indication or judgment that the supplier of a product or service has met the requirements of the relevant specifications, contract, or regulation; also the state of meeting the requirements. [Von Zharen, ISO 14000, p. 194 (1996)]

Compliance Schedule (合规时间表) (hé guī shí jiān biǎo)

A schedule of remedial measures included in a permit or an enforcement order, including a sequence of interim requirements (for example, actions, operations, or milestone events) that lead to compliance with the CWA and regulations. [US EPA Glossary of NPDES Terms (2004)]

Composite Sample (混合样品) (hùn hé yàng pǐn)

Sample composed of two or more discrete samples. The aggregate sample will reflect the average water quality covering the compositing or sample period. [US EPA Glossary of NPDES Terms (2004)]

Composting (堆肥) (duī féi)

Managing the decomposition of organic materials, such as leaves, grass, and garden waste. [Urban Soil Primer, p. 71 (2005)]

Comprehensive Environmental Response, Compensation, and Liability Act (CERCLA, United States) (综合环境反应、赔偿和责任法 (美国)) (zòng hé huán jìng fǎn yìng, péi cháng hé zé rèn fǎ (měi guó))

A major US federal law enacted in 1980, CERCLA created a tax on the chemical and petroleum industries and provided broad Federal authority to respond directly to releases or threatened releases of hazardous substances that may endanger public health or the environment. [US EPA, CERCLA (2012)]

Concentrated Animal Feeding Operation (CAFO) (动物集中饲养作业) (dòng wù jí zhōng sì yǎng zuò yè)

An AFO that is defined as a Large CAFO or as a Medium CAFO ... or that is designated as a CAFO ... Two or more AFOs under common ownership are considered to be a single AFO for the purposes of determining the number of animals at an operation, if they adjoin each other or if they use a common area or system for the disposal of wastes. [USEPA Glossary of NPDES Terms (2004)] See **Large Concentrated Animal Feeding Operation** and **Medium Concentrated Animal Feeding Operation**.

Concentration (浓度) (nóng dù)

Amount of a chemical in a particular volume or weight of air, water, soil, or other medium. [US EPA Glossary of Climate Change Terms (2011)]

Concessional and Preferential Terms (减让和优惠条款) (jiǎn ràng hé yōu huì tiáo kuǎn)

Provisions or terms that are expressly included in an agreement to induce a certain party to assume the agreement's commitments or obligations, without which this party would not agree to enter into the agreement.

Conclusive Evidence (确凿证据) (què záo zhèng jù)

Facts that can be independently verified to establish beyond a reasonable doubt that the facts do exist as reported; evidence that cannot be contradicted.

Conditioned Air (经调节的空气) (jīng tiáo jié de kōng qì)

Air that has been heated, cooled, humidified, or dehumidified to maintain an interior space within the "comfort zone" (sometimes referred to as "tempered" air). [US EPA Glossary of Indoor Air Quality Terms (2011)]

Conference of the Parties (缔约方会议) (dì yuē fāng huì yì)

(1) The supreme body of the United Nations Framework Convention on Climate Change (UNFCCC). It comprises more than 180 nations that have ratified the Convention. Its first session was held in Berlin, Germany, in 1995 and it is expected to continue meeting on a yearly basis. The COP's role is to promote and review the implementation of the Convention. It will periodically review existing commitments in light of the Convention's objective, new scientific findings, and the effectiveness of national climate change programs. [US EPA Glossary of Climate Change Terms (2011)] (2) The plenary assembly of the States that are the contracting parties to a multilateral environmental agreement or other treaty, with such powers as provided in the international agreement that constitutes the COP.

Conformance (合格) (hé gé)

An affirmative indication or judgment that a product or service has met the requirements of the relevant specifications, contract, or regulation; also the state of meeting the requirements. [Von Zharen, ISO 14000, p. 194 (1996)]

Conformity Assessment (合格评估) (hé gé píng gū)

Conformity assessment includes all activities that are intended to assure the conformity of products to a set of standards. This can include testing, inspection, certification, quality system assessment, and other activities. [Von Zharen, ISO 14000, p. 195 (1996)]

Constant Air Volume Systems (恒风量系统) (héng fēng liàng xì tǒng)

Air handling system that provides a constant air flow while varying the temperature to meet heating and cooling needs. [US EPA Glossary of Indoor Air Quality Terms (2011)]

Consultative Advice (咨询建议) (zī xún jiàn yì)

Technical or other expert information and assessments provided by an independent authority that is requested to provide an evaluation of a specified matter.

Consumption (消费量) (xiāo fèi liàng)

Production plus imports minus exports of controlled substances. [Montreal Protocol, Art. 1, No.6 (1987)]

Container Collection and Classification of Packaging Recycling Law (容器包装物的分类收集与循环法) (róng qì bāo zhuāng wù de fēn lèi shōu jí yǔ xún huán fǎ)

A Japanese law promoting the conversion of container and wrapping waste into a resource.

Container Gardens (盆栽花园) (pén zāi huā yuán)

Gardens planted in pots, concrete boxes, brick, or stone basins, or other isolated rooting areas within paved areas. [Urban Soil Primer, p. 71 (2005)] Container gardens are distinguished from gardens planted directly into the surface of the ground. Such gardens allow apartment dwellers and other urban residents without land to grow plants on balconies, sidewalks or other surfaces, and allow gardeners with land to arrange their plants artistically or raise them above ground level for practical reasons such as drainage or pest avoidance.

Contaminated Soil (污染的土壤) (wū rǎn de tǔ rǎng)

A soil that has high concentrations of trace metals or organic waste that is toxic or a high risk to people or animals. [Urban Soil Primer, p. 71 (2005)]

Contamination (污染) (wū rǎn)

1. Radioactive substances on surfaces, or within solids, liquids or gases (including the human body), where their presence is unintended or undesirable, or the process giving rise to their presence in such places. Also used less formally to refer to a quantity, namely the activity on a surface (or on a unit area of a surface). Contamination does not include residual radioactive material remaining at a site after the completion of decommissioning. The term contamination may have a connotation that is not intended. The term contamination refers only to the presence of radioactivity, and gives no indication of the magnitude of the hazard involved. 2. The presence of a radioactive substance on a surface in quantities in excess of 0.4 Bq/cm^2 for beta and gamma emitters and low toxicity alpha emitters, or 0.04 Bq/cm^2 for all other alpha emitters. (From Ref. [2]) (This is a regulatory definition of contamination, specific to the Transport Regulations. Levels below 0.4 Bq/cm^2 or 0.04 Bq/cm^2 would still be considered contamination according to the scientific definition (1).) [IAEA Safety Glossary, p. 41

Continental Shelf (大陆架) (dà lù jià)

The continental shelf of a coastal State comprises the seabed and subsoil of the submarine areas that extend beyond its territorial sea throughout the natural prolongation of its land territory to the outer edge of the continental margin, or to a distance of 200 nautical miles from the baselines from which the breadth of the territorial sea is measured where the outer edge of the continental margin does not extend up to that distance. [Law of the Sea Convention, Art. 76 (1982)]

Contingency Plans Against Pollution (污染应急计划) (wū rǎn yìng jí jì huà)

(1) A contingency plan, in general, is a plan that addresses a possible, though not expected, occurrence. (2) In the cases referred to in article 198, States in the area affected, in accordance with their capabilities, and the competent international organizations shall cooperate, to the extent possible, in eliminating the effects of pollution and preventing or minimizing the damage. To this end, States shall jointly develop and promote contingency plans for responding to pollution incidents in the marine environment. [Law of the Sea Convention, Art.199 (1982)]

Continual Improvement (持续改进) (chí xù gǎi jìn)

Process of enhancing the environmental management system, with the purpose of achieving improvements in overall environmental perform- ance, not necessarily in all areas of activity simultaneously, resulting from continuous efforts to improve in line with the organization's environmen- tal policy. [Von Zharen, ISO 14000, p. 195 (1996)]

Contracting Parties (缔约当事方) (dì yuē dāng shì fāng)

The sovereign States that formally agree to sign and ratify or adhere to an international agreement; also used to refer to authorities or entities or individuals who agree to a contract or other written agreement.

Contractor (订约人) (dìng yuē rén)

The organization that provides a product to the customer in a contractual situation. [Von Zharen, ISO 14000, p. 195 (1996)]

Control of Weapons of Mass Destruction (大规模毁灭性武器管制) (dà guī mó huǐ miè xìng wǔ qì guǎn zhì)

(1) In security and foreign policy analyses, "weapons of mass destruc- tion" (WMD) is a term that generally encompasses nuclear, chemical, and biological weapons, with radiological weapons occasionally included. Contemporary international legal analysis generally follows this conven- tional definition of WMD, even though neither treaty law nor customary international law contains an authoritative definition of WMD. [ASIL Insights] (2) The measures approved by the United Nations Security Council, or by international agreements, such as the 1963 convention banning the atmospheric testing of nuclear weapons, or in accord with the resolutions of the United Nation General Assembly, to restrict or elimi- nate the use of WMD.

Controlled Substance (控制物质) (kòng zhì wù zhì)

A substance in Annex A, Annex B, Annex C, or Annex E to this Protocol, whether existing alone or in a mixture. It includes the isomers of any such substance, except as specified in the relevant Annex, but excludes any controlled substance or mixture which is in a manufactured product other than a container used for the transportation or storage of that substance. [Montreal Protocol, Art. 1, No. (1987)]

Convention (1) (公约) (gōng yuē)

A treaty or formal agreement between nations.

Convention (2) (惯例) (guàn lì)

A customary practice, rule or method. [Von Zharen, ISO 14000, p. 195 (1996)]

Convention Concerning Safety in the Use of Chemicals at Work (化学制品在工作中的使用安全公约) (huà xué zhì pǐn zài gōng zuò zhōng de shǐ yòng ān quán gōng yuē)

An International Labour Organisation (ILO) Convention on chemical safety in the workplace adopted June 25, 1990, and entered into force November 4, 1993. "Safety in the use of chemicals at work" is also the title of a 1993 ILO code of practice, which provides practical guidance to carry out the Convention and is a contribution to the International Programme on Chemical Safety of the United Nations Environment Programme, the International Labour Organisation and the World Health Organisation.

Convention Concerning the Protection of the World Cultural And Natural Heritage (保护世界文化和自然遗产公约) (bǎo hù shì jiè wén huà hé zì rán yí chǎn gōng yuē)

The United Nations Educational, Scientific, and Cultural Organization formed this Convention to address threats to cultural heritage and natural heritage. The treaty was signed on November 16, 1972.

Convention for the Conservation of Antarctic Seals (保护南极海豹公约) (bǎo hù nán jí hǎi bào gōng yuē)

Entered into force in 1978, this Convention aims to promote and achieve the protection, scientific study and rational use of Antarctic seals, and to maintain a satisfactory balance within the ecological system of the Antarctic.

Convention for the Prevention of Marine Pollution from Land-based Sources (防止陆源污染海洋公约) (fáng zhǐ lù yuán wū rǎn hǎi yáng gōng yuē)

Also called the Paris Convention, entered into force in 1978. [Convention on Marine Pollution (1974)]

Convention for the Protection of the Marine Environment of the North-east Atlantic (OSPAR Convention) (保护东北大西洋海洋环境公约) (bǎo hù dōng běi dà xī yáng hǎi yáng huán jìng gōng yuē)

A regional marine environmental treaty opened for signature on September 22, 1992 and entered into force March 25, 1998. This treaty aims to protect the Northeast Atlantic marine environment against pollution caused by human activity.

Convention for the Safeguarding of the Intangible Cultural Heritage (保护非物质文化遗产公约) (bǎo hù fēi wù zhì wén huà yí chǎn gōng yuē)

A Convention that commits states to ensure the safeguarding of intangible cultural heritage present in their territories, adopted by the General Conference of UNESCO on October 17, 2003.

Convention on Biological Diversity (生物多样性公约) (shēng wù duō yàng xìng gōng yuē)

A 1992 convention that entered into force on December 29, 1993. It has three main objectives: (1) the conservation of biological diversity; (2) the sustainable use of the components of biological diversity; and (3) the fair and equitable sharing of the benefits arising out of the utilization of genetic resources. [Convention on Biological Diversity (1992)]

Convention on Early Notification of a Nuclear Accident (核事故及早通报公约) (hé shì gù jí zǎo tōng bào gōng yuē)

Convention of the International Atomic Energy Agency that was adopted in 1986 following the Chernobyl nuclear plant accident. [Convention on Assistance in the Case of a Nuclear Accident (1987)]

Convention on International Liability for Damage Caused by Space Objects (国际空间实体赔偿责任公约) (guó jì kōng jiān shí tǐ péi cháng zé rèn gōng yuē)

Convention entered into force in September 1972. It provides that a launching State shall be absolutely liable to pay compensation caused by its space objects on the surface of the Earth or to aircraft, and liable for damage due to its fault in space. [UNOOSA]

Convention on International Trade in Endangered Species of Wild Fauna and Flora (CITES) (濒危野生动植物物种国际贸易公约) (bīn wēi yě shēng dòng zhí wù wù zhǒng guó jì mào yì gōng yuē)

An international convention that aims to ensure that the survival of wild animals and plants is not threatened by international trade in specimens. Entered into force July 1, 1975.

Convention on Nuclear Safety (核安全公约) (hé ān quán gōng yuē)

Adopted on June 17, 1994, in Vienna, this convention aims to legally commit participating States operating land-based nuclear power plants to maintain a high level of safety by setting international benchmarks to which States would subscribe. [Convention on Nuclear Safety]

Convention on Protection of the Rhine against Chloride Pollution (保护莱茵河不受氯化物污染公约) (bǎo hù lái yīn hé bú shòu lù huà wù wū rǎn gōng yuē)

A multilateral treaty, entered into force July 5, 1985. The Convention's objective is to protect the Rhine against chloride pollution with the purpose of ameliorating water standards.

Convention on Registration of Objects Launched into Outer Space (发射外层空间物体登记公约) (fā shè wài céng kōng jiān wù tǐ dēng jì gōng yuē)

Entered into force in 1976, this Convention aims to detail the orbits of objects launched into outer space.

Convention on the Conservation of Antarctic Marine Living Resources (南极海洋生物资源养护公约) (nán jí hǎi yáng shēng wù zī yuán yǎng hù gōng yuē)

Convention to preserve marine life and environmental integrity in and near Antarctica, entered into force in 1982 as part of the Antarctic Treaty System.

Convention on the Conservation of Migratory Species of Wild Animals (保护野生动物迁徙物种公约) (bǎo hù yě shēng dòng wù qiān xī wù zhǒng gōng yuē)

Also known as the Bonn Convention, this convention aims to conserve terrestrial, aquatic, and avian migratory species throughout their range; it entered into force in 1983.

Convention on the Means Prohibiting and Preventing the Illicit Import, Export and Transfer of Ownership of Cultural Property (关于禁止和防止非法进口、出口和文化财产所有转让的方法的公约) (guān yú jìn zhǐ hé fáng zhǐ fēi fǎ jìn kǒu 、 chū kǒu hé wén huà cái chǎn suǒ yǒu zhuǎn ràng de fāng fǎ de gōng yuē)

A treaty opened for signature on November 14, 1970, this agreement aims to curb the illicit import, export, and transfer of ownership of cultural property. Entered into force April 24, 1972.

Convention on the Physical Protection of Nuclear Material (核材料实体保护公约) (hé cái liào shí tǐ bǎo hù gōng yuē)

Entered into force February 8, 1987, this Convention establishes measures related to the prevention, detection, and punishment of offenses relating to nuclear material.

Convention on the Prevention of Marine Pollution by Dumping of Wastes and Other Matter (防止倾倒废物及其他物质污染海洋公约) (fáng zhǐ qīng dǎo fèi wù jí qí tā wù zhì wū rǎn hǎi yáng gōng yuē)

Also called the London Convention, this treaty entered into force in 2006. The Protocol, which will eventually replace the 1972 Convention, represents a major change of approach to the question of how to regulate the use of the sea as a depository for waste materials. Article 3 stresses the "precautionary approach." This requires that appropriate preventive measures are taken when there is reason to believe that wastes or other matter introduced into the marine environment are likely to cause harm even when there is no conclusive evidence to prove a causal relation between inputs and their effects. The article also states that "the polluter should, in principle, bear the cost of pollution" and it emphasizes that Contracting Parties should ensure that the Protocol should not simply result in pollution being transferred from one part of the environment to another. Rather than state which materials may not be dumped, the 1996 Protocol restricts all dumping except for a permitted list. [Convention on the Prevention of Marine Pollution by Dumping of Waste and Other Matter (2006)]

Convention on the Prohibition of Military or Any Other Hostile Use of Environmental Modification Techniques (关于禁止军事或为任何其他敌对目的使用改变环境技术的公约) (guān yú jìn zhǐ jūn shì huò wéi rèn hé qí tā dí duì mù de shǐ yòng gǎi biàn huán jìng jì shù de gōng yuē)

Also known as the Environmental Modification Convention, this treaty prohibits the military or other hostile use of environmental modification techniques. This treaty was signed on May 18, 1977, and entered into force on October 5, 1978.

Convention on the Protection and Use of Transboundary Watercourses and International Lakes (1992 Helsinki Convention) (跨界水体和国际湖泊利用和保护公约 (1992年的赫尔辛基公约)) (kuà jiè shuǐ tǐ hé guó jì hú pō lì yòng hé bǎo hù gōng yuē (1992 nián hè ěr xīn jī gōng yuē))

A treaty signed on March 17, 1992 and entered into force 1998. The treaty aims to prevent, control, and reduce transboundary impacts on international lakes and watercourses.

Convention on Transboundary Effects of Industrial Accidents (工业事故跨国影响公约) (gōng yè shì gù kuà guó yǐng xiǎng gōng yuē)

This 1992 Convention, which entered into force April 19, 2000, aims to protect people and the environment against industrial accidents.

Convention on Wetlands of International Importance (关于特别是作为水禽栖息地的国际重要湿地公约) (guān yú tè bié shì zuò wéi shuǐ qín qī xī dì de guó jì zhòng yào shī dì gōng yuē)

Also called the Ramsar Convention, entered into force in 1975. A treaty to address wetland habitat loss and degradation.

Conventional Pollutants (常规污染物) (cháng guī wū rǎn wù)

Pollutants typical of municipal sewage, and for which municipal secondary treatment plants are typically designed; defined by Federal Regulation [40 CFR 401.16] as BOD, TSS, fecal coliform bacteria, oil, and grease, and pH. [US EPA Glossary of NPDES Terms (2004)]

Conventional Septic System (常规化粪池) (cháng guī huà fèn chí)

A wastewater treatment system consisting of a septic tank and a typical trench or bed subsurface wastewater infiltration system. [US EPA Glossary of NPDES Terms (2004)]

Corfu Channel Case (United Kingdom v. Albania) (ICJ, 1949) (英国诉阿尔巴尼亚科孚海峡案 (国际法院, 1949年)) (yīng guó sù ā'ěr bā ní yà kē fú hǎi xiá àn (guó jì fǎ yuàn, 1949 nián))

A 1949 decision by the International Court of Justice regarding the United Kingdom's claim that Albania caused or had knowledge of the placing of mines in its territorial waters in the Strait of Corfu without notifying parties of the existence of these mines and thus violating the Hague Convention. [Corfu Channel (1947)]

Corrective Action (校正行动) (jiào zhèng xíng dòng)

An action taken to eliminate the causes of an existing nonconformity, defect or other undesirable situation in order to prevent recurrence. [Von Zharen, ISO 14000, p. 195 (1996)]

Countries in Transition to Market Economies (向市场经济过渡的国家) (xiàng shì chǎng jīng jì guò dù dē guó jiā)

The nations of the former Union of Soviet Socialist Republics and States of Central and Eastern Europe whose economies were directed by the central planning systems of their former Communist economic orders, and whose governments are introducing liberal economic market systems for their commercial transactions and financing.

Country of Origin (原产国) (yuán chǎn guó)

The country from where the genetic resources and/or biological material originated. A country of origin can be a primary or secondary center of origin for the material. [UNU IAS Pocket Guide, p. 12 (2007)]

Country of Origin of Genetic Resources (遗传资源原产国) (yí chuán zī yuán yuán chǎn guó)

The country which possesses those genetic resources in *in-situ* conditions. [Convention on Biological Diversity, Art. 2 (1992)]

Country Providing Genetic Resources (遗传资源提供国) (yí chuán zī yuán tí gòng guó)

The country supplying genetic resources collected from *in-situ* sources, including populations of both wild and domesticated species, or taken from *ex-situ* sources, which may or may not have originated in that country. [Convention on Biological Diversity, Art. 2 (1992); UNU IAS Pocket Guide, p. 12 (2007)]

Criteria (标准) (biāo zhǔn)

The numeric values and the narrative standards that represent contaminant concentrations that are not to be exceeded in the receiving environmental media (surface water, ground water, sediment) to protect beneficial uses. [US EPA Glossary of NPDES Terms (2004)]

Cryosphere (冰冻圈) (bīng dòng quān)

One of the interrelated components of the Earth's system, the cryosphere is frozen water in the form of snow, permanently frozen ground (permafrost), floating ice, and glaciers. Fluctuations in the volume of the cryosphere cause changes in ocean sea level, which directly impact the atmosphere and biosphere. [US EPA Glossary of Climate Change Terms (2011)]

Cultural Heritage (文化遗产) (wén huà yí chǎn)

Both tangible pieces of heritage – palaces, temples, and other historic landmarks – and intangible heritage, including acts of creation, representation, and processes of transmission, such as the performing arts, languages, and oral traditions. [UNEP Judicial Handbook, Ch. 12.1 (2005)]

Culture (文化) (wén huà)

The totality of socially transmitted behavior patterns, arts, beliefs, institutions, and all other products of human work and thought. [UNEP Judicial Handbook, Ch. 12.1 (2005)]

Customer (消费者) (xiāo fèi zhě)

Ultimate consumer, user, client, beneficiary, or second party. [Von Zharen, ISO 14000, p. 195 (1996)]

CWA

See **Clean Water Act**.

D

Daily Discharge (日排放) (rì pái fàng)

The discharge of a pollutant measured during any 24-hour period that reasonably represents a calendar day for purposes of sampling. For pollutants with limitations expressed in units of mass, the daily discharge is calculated as the total mass of the pollutant discharged during the day. For pollutants with limitations expressed in other units of measurement (e.g., concentration) the daily discharge is calculated as the average measurement of the pollutant throughout the day (40 CFR 122.2). [US EPA Glossary of NPDES Terms (2004)]

Daily Maximum Limit (日最大限度) (rì zuì dà xiàn dù)

The maximum allowable discharge of pollutant during a calendar day. Where daily maximum limitations are expressed in units of mass, the daily discharge is the total mass discharged over the course of the day. Where daily maximum limitations are expressed in terms of a concentration, the daily discharge is the arithmetic average measurement of the pollutant concentration derived from all measurements taken that day. [US EPA Glossary of NPDES Terms (2004)]

Dampers (气流调节器) (qì liú tiáo jiē qì)

Controls that vary airflow through an air outlet, inlet, or duct. A damper position may be immovable, manually adjustable or part of an automated control system. [US EPA Glossary of Indoor Air Quality Terms (2011)]

Data Quality (数据质量) (shù jù zhì liàng)

Degree of confidence in individual input data from a source, aggregated data and in the data set as a whole. Life Cycle Assessment data quality is described by established indicators selected for the study. [Von Zharen, ISO 14000, p. 195 (1996)]

Decision (决定) (jué dìng)

The final determination of a local reviewing body or administrative agency or officer regarding an application for a permit or approval. [Nolon, Well Grounded, p. 448 (ELI 2001)]

Declaration on the Human Environment (人类环境宣言) (rén lèi huán jìng xuān yán)

Declaration from the meeting in Stockholm June 5–16, 1972. This Declaration addressed the "need for a common outlook and for common principles to inspire and guide the peoples of the world in the preservation and enhancement of the human environment." [Declaration on the Human Environment]

Declaration on the Protection of the Arctic Environment (北冰洋环境保护宣言) (běi bīng yáng huán jìng bǎo hù xuān yán)

Also called the Rovaniemi Declaration, this declaration was signed by Canada, Denmark, Finland, Iceland, Norway, Sweden, the USSR, and the United States on June 14, 1991. Its objectives are "[P]reserving environmental quality and natural resources, accommodating environmental protection principles with the needs and traditions of Arctic Native peoples, monitoring environmental conditions, and reducing and eventually eliminating pollution in the Arctic Environment."

Decontamination (净化) (jìng huà)

The complete or partial removal of contamination by a deliberate physical, chemical, or biological process. This definition is intended to include a wide range of processes for removing contamination from people, equipment and buildings, but to exclude the removal of radionuclides from within the human body or the removal of radionuclides by natural weathering or migration processes, which are not considered to be decontamination. [IAEA Safety Glossary, p. 48–9 (2007)] See **Remediation**.

Decontamination Factor (净化系数) (jìng huà xì shù)

The ratio of the activity per unit area (or per unit mass or volume) before a particular decontamination technique is applied to the activity per unit area (or per unit mass or volume) after application of the technique. This ratio may be specified for a particular radionuclide or for gross activity. The background activity may be first deducted from the activity per unit area both before and after a particular decontamination technique is applied. [IAEA Safety Glossary, p. 49 (2007)]

Deed Restrictions (契据限制) (qì jù xiàn zhì)

A covenant or restriction placed in a deed that restricts the use of the land in some way. These are often used to insure that the owner complies with a condition imposed by a land use body. [Nolon, Well Grounded, p. 448 (ELI 2001)]

Deforestation (采伐森林) (cǎi fá sēn lín)

Those practices or processes that result in the conversion of forested lands for non-forest uses. This is often cited as one of the major causes of the enhanced greenhouse effect for two reasons: (1) the burning or decomposition of the wood releases carbon dioxide; and (2) trees that once removed carbon dioxide from the atmosphere in the process of photosynthesis are no longer present. [US EPA Glossary of Climate Change Terms (2011)]

Deliberate Disposal (故意处置) (gù yì chǔ zhì)

(1) Dumping or disposal of wastes or man-made structures into the sea. [Law of the Sea Convention, Art. 1, § 5 (a) (1982)] (2) The knowing act of dumping or discharging waste or other substances, in order to discard or remove it from the control of whomever has had custody of the waste or substances.

Density (密度) (mì dù)

The amount of development permitted per acre on a parcel under the zoning law. The density allowed could be, for example, four dwelling units per acre or 40 000 square feet of commercial building floor per acre. [Nolon, Well Grounded, p. 448 (ELI 2001)]

Density Bonus (密度奖励) (mì dù jiǎng lì)

See **Incentive Zoning**.

Desert (荒漠) (huāng mò)

A land area so dry that little or no plant or animal life can survive. [NASA Earth Observatory (2012)]

Desertification (荒漠化) (huāng mò huà)

(1) Land degradation in arid, semi-arid, and dry sub-humid areas resulting from various factors, including climatic variations and human activities. [Convention to Combat Desertification, Art. 1 (1994); UNEP Judicial Handbook, 9.3.1 (2005)] (2) Land degradation in arid, semi-arid, and dry sub-humid areas resulting from various factors, including climatic variations and human activities. Further, the UNCCD (The United Nations Convention to Combat Desertification) defines land degradation as a reduction or loss, in arid, semi-arid, and dry sub-humid areas, of the biological or economic productivity and complexity of rain-fed cropland, irrigated cropland, or range, pasture, forest, and woodlands resulting from land uses or from a process or combination of processes, including processes arising from human activities and habitation patterns, such as: (i) soil erosion caused by wind and/or water; (ii) deterioration of the physical, chemical, and biological or economic properties of soil; and (iii) long-term loss of natural vegetation. Conversion of forest to non-forest. [US EPA Glossary of Climate Change Terms (2011)]

Design Review (设计审查) (shè jì shěn chá)

A formal, documented, comprehensive, and systematic examination of a design to evaluate the design requirements and the capability of the design to meet these requirements and to identify problems and propose solutions. [Von Zharen, ISO 14000, p. 196 (1996)]

Destruction of Hazardous Constituents (破坏有害成分) (pò huài yǒu hài chéng fèn)

The processes or action required to transform chemical or other characteristics of a substance from a condition in which the constituents are harmful to life or constitute a hazard to a condition in which they are inert or neutral with respect to the actual or potential adverse impacts posed.

Detailed Description (详尽陈述) (xiáng jìn chén shù)

A written statement providing facts and analysis sufficient to permit a reader to gain a clear understanding of the matter set forth in the statement. In the case of an environmental impact assessment the detailed written statement should be written clearly enough for any member of the public to understand the matters being described.

Determination (判定) (pàn dìng)

A decision rendered by an officer or administrative body on an application or a request for a ruling. [Nolon, Well Grounded, p. 448 (ELI 2001)]

Developed Country Parties (发达国家缔约方) (fā dá guó jiā dì yuē fāng)

Under the Desertification Convention, includes developed country Parties and regional economic integration organizations constituted by developed countries. [Convention to Combat Desertification, Art. 1 (1994)]

Development Document (编制文件) (biān zhì wén jiàn)

A report prepared during the development of an effluent limitation guideline by EPA that provides the data and methodology used to develop limitations guidelines and categorical pretreatment standards for an industrial category. [US EPA Glossary of NPDES Terms (2004)]

Development Overlay Zones (发展叠加区) (fā zhǎn dié jiā qū)

In development overlay zones, the legislature may provide incentives, such as density bonuses or waivers of certain zoning requirements, for developers who build the type of development desired. [Nolon, Well Grounded, p. 448 (ELI 2001)]

Diffusers and Grilles (扩散器和格栅) (kuò sàn qì hé gé shān)

Components of the ventilation system that distribute and return air to promote air circulation in the occupied space. As used in this document, supply air enters a space through a diffuser or vent and return air leaves a space through a grille. [US EPA Glossary of Indoor Air Quality Terms (2011)]

Directly or Indirectly (直接或间接地) (zhí jiē huò jiàn jiē dī)

The method whereby an action causes an effect, either by being the proximate or immediate cause of the action in the case of a direct act, or by being the secondary or remote cause of the action, in the case of an indirect act.

Director (主管) (zhǔ guǎn)

The Regional Administrator or State Director, as the context requires, or an authorized representative. When there is no approved state program, and there is an EPA administered program, Director means the Regional Administrator. When there is an approved state program, "Director" normally means the State Director. [US EPA Glossary of NPDES Terms (2004)]

Disadvantaged People (弱势群体) (ruò shì qún tǐ)

(Agricultural) A person or group of people engaged in, or who rely on agriculture and who, for various reasons, may be deprived of a reasonable standard of living, access to information, knowledge, health, education, opportunities, access to markets, and other services and benefits, by virtue

of their socio-economic, cultural, ethnic, or locational circumstances.
[Drafting Legislation for Sustainable Soils: A Guide (2004)]

Discharge (排放) (pái fàng)

1. Planned and controlled release of (usually gaseous or liquid) radioactive
material to the environment. Strictly, the act or process of releasing the
material, but also used to describe the material released.

Authorized discharge: (经授权的排放) (jīng shòu quán dě pái fàng)
Discharge in accordance with an authorization.

Radioactive discharge: (放射性物质排放) (fàng shè xìng wù zhì pái fàng)
Radioactive substances arising from a source within a practice which are
discharged as gases, aerosols, liquids, or solids to the environment, generally with the purpose of dilution and dispersion.

2. A planned and controlled release to the environment, as a legitimate
practice, within limits authorized by the regulatory body, of liquid or
gaseous radioactive material that originate from regulated nuclear facilities during normal operation. [IAEA Safety Glossary, p. 53 (2007)]

Discharge Monitoring Report (DMR) (排放监测报告) (pái fàng jiān cè bào gào)

The form used (including any subsequent additions, revisions, or modifi-
cations) to report self-monitoring results by NPDES permittees. DMRs
must be used by approved states as well as by EPA. [US EPA Glossary of
NPDES Terms (2004)]

Disinfectants (消毒剂) (xiāo dú jì)

One of three groups of antimicrobials registered by EPA for public-health
uses. EPA considers an antimicrobial to be a disinfectant when it destroys
or irreversibly inactivates infectious or other undesirable organisms, but
not necessarily their spores. EPA registers three types of disinfectant
products based upon submitted efficacy data: limited, general, or broad
spectrum, and hospital disinfectant. [US EPA Glossary of Indoor Air
Quality Terms (2011)]

Dispersal (扩散) (kuò sàn)

The spreading of radioactive material in the environment. In normal language synonymous with dispersion, but tends to be used in a general sense, not implying the involvement of any particular process or phenomenon, e.g., the uncontrolled spreading of material that has escaped from confinement, or as a result of damage to (or the destruction of) a sealed source, special form radioactive material or low dispersible radioactive material. [IAEA Safety Glossary, p. 54 (2007)]

Disposer (处置者) (chǔ zhì zhě)

Any person to whom hazardous wastes or other wastes are shipped and who carries out the disposal of such wastes. [Basel Convention, Art. 2, No. 19 (1989)]

District (区) (qū)

A portion of a community identified on the locality's zoning map within which one or more principal land uses are permitted along with their accessory uses and any special land uses permitted by the zoning provisions for the district. [Nolon, Well Grounded, p. 448 (ELI 2001)]

Dolphin (海豚) (hǎi tún)

Coryphaena hippurus; *Coryphaena equiselis* [Law of the Sea Convention, Annex I (1982)]

Domesticated or Cultivated Species (驯化或培育物种) (xùn huà huò péi yù wù zhǒng)

Species in which the evolutionary process has been influenced by humans to meet their [humans'] needs. [Convention on Biological Diversity, Art. 2 (1992); UNEP Guide to Climate Neutrality, p. 12 (2008)]

Downwelling (沉降流) (chén jiàng liú)

The process of accumulation and sinking of warm surface waters along a coastline. A change of air flow of the atmosphere can result in the sinking or downwelling of warm surface water. The resulting reduced nutrient supply near the surface affects the ocean productivity and meteorological conditions of the coastal regions in the downwelling area. [NASA Earth Observatory (2012)]

Draft Permit (许可证初步审批文件) (xǔ kě zhèng chū bù shěn pī wén jiàn)

A document prepared under 40 CFR 124.6 indicating the Director's tentative decision to issue, deny, modify, revoke and reissue, terminate, or reissue a permit. A notice of intent to terminate a permit, and a notice of intent to deny a permit application, as discussed in 40 CFR 124.5, are considered draft permits. A denial of a request for modification, revocation and reissuance, or termination, as discussed in 40 CFR 124.5, is not a draft permit. [US EPA Glossary of NPDES Terms (2004)]

Drain Tile Loop (排水瓦管环路) (pái shuǐ wǎ guǎn huán lù)

A continuous length of drain tile or perforated pipe extending around all or part of the internal or external perimeter of a basement or crawlspace footing. [US EPA Glossary of Indoor Air Quality Terms (2011)]

Drain Trap (排水阱) (pái shuǐ jǐng)

A dip in the drain pipe of sinks, toilets, floor drains, etc., which is designed to stay filled with water, thereby preventing sewer gases from escaping into the room. [US EPA Glossary of Indoor Air Quality Terms (2011)]

Dredged Material (疏浚物) (shū jùn wù)

(1) Substances that are removed by excavation or otherwise from having been located under water or in sediments or in wetlands; (2) The removal of material from the bottom of lakes, rivers, harbors and other water bodies. Most dredging is done to maintain or deepen navigation channels,

anchorages or berthing areas for the safe passage of boats and ships. Dredging of contaminated areas may also be performed for the express purpose of reducing the exposure of marine biota and humans to contaminants and/or to prevent the spread of contaminants to other areas of the water body. This type of dredging is termed "environmental dredging." The increasing need for wider and deeper channels and different placement methods has spawned the development of the many types of dredges. [EPA Region 2]

Drought (干旱) (gān hàn)

The naturally occurring phenomenon that exists when precipitation has been significantly below normal recorded levels, causing serious hydrological imbalances that adversely affect land resource production systems. [Convention to Combat Desertification, Art. 1 (1994)]

Dump Site (倾倒地点) (qīng dào dì diàn)

The location where wastes are discarded or disposed of or taken for storage. See **Landfill**.

Dumping (倾倒) (qīng dào)

(1) Any deliberate disposal of wastes or other matter from vessels, aircraft, platforms or other man-made structures at sea; (2) any deliberate disposal of vessels, aircraft, platforms or other man-made structures at sea; "dumping" does not include: (i) the disposal of wastes or other matter incidental to, or derived from the normal operations of vessels, aircraft, platforms or other man-made structures at sea and their equipment, other than wastes or other matter transported by or to vessels, aircraft, platforms or other man-made structures at sea, operating for the purpose of disposal of such matter or derived from the treatment of such wastes or other matter on such vessels, aircraft, platforms or structures; (ii) placement of matter for a purpose other than the mere disposal thereof, provided that such placement is not contrary to the aims of this Convention. [Law of the Sea Convention, Art. 1 (1982)]

Duty of Care (注意义务) (zhù yì yì wù)

The obligation that one person or party owes toward another person or party because of the relationship between them whereby the act or omission to act by the first person or party would have an adverse impact on the second persons or parties. This relationship gives rise to a duty on the part of the first person or party to take reasonable care to recognize and avoid the potential harm that may be caused to the other persons or parties.

Dwelling Unit (住宅单位) (zhù zhái dān wèi)

A unit of housing with full housekeeping facilities for a family. [Nolon, Well Grounded, p. 448 (ELI 2001)]

E

E Horizon (E层, 淋溶层) (E céng, lín róng céng)

The mineral horizon in which the main feature is loss of silicate clay, iron, aluminum, or some combination of these. [Urban Soil Primer, p. 72 (2005)]

Easement (地役权) (dì yì quán)

An easement involves the right to use a parcel of land to benefit an adjacent parcel of land, such as to provide vehicular or pedestrian access to a road or sidewalk. Technically known as an easement appurtenant. [Nolon, Well Grounded, p. 448 (ELI 2001)]

Eccentricity (偏心率 (地球)) (piān xīn lǜ (dì qiú))

The extent to which the Earth's orbit around the Sun departs from a perfect circle. [US EPA Glossary of Climate Change Terms (2011)]

Eco-driving (生态-驾驶) (shēng tài - jià shǐ)

Eco-driving is a way of driving that reduces fuel consumption, greenhouse gas emissions, and accident rates. [UNEP Guide to Climate Neutrality, p. 194 (2008)]

Ecological Function (生态功能) (shēng tài gōng néng)

The natural processes, products, or services that living and non-living environments provide or perform within or between species, ecosystems, and landscapes. These may include biological, physical, and socio-economic interactions [OT Ministry of Municipal Affairs & Housing Land Use Planning Definitions (2007)]; a biological or geo-chemical interaction within a natural system that maintains the balance of other natural and physical functions in that ecosystem.

Ecological Integrity of Soil (土壤生态完整性) (tǔ rǎng shēng tài wán zhěng xìng)

Preserving the wholeness of ecosystems, including the prevention of loss of wholeness, so as to stop the commencement of soil degradation, and to control existing soil degradation, and to protect and manage soil for its sustainable use. [Drafting Legislation for Sustainable Soils: A Guide, p. 71 (2004)] See **Ecologically Sustainable**.

Ecological Soil Standard (生态土壤标准) (shēng tài tǔ rǎng biāo zhǔn)

A process for maintaining or improving the ecological integrity of soil. [Drafting Legislation for Sustainable Soils: A Guide, p. 72 (2004)]

Ecologically Sustainable (生态可持续的) (shēng tài kě chí xù de)

(In soil standards): The sustained functioning of ecosystems by using appropriate ecological soil standards. [Drafting Legislation for Sustainable Soils: A Guide, p. 91 (2004)]

Ecology (生态学 (shēng tài xué)

The study of natural systems and their functions, such as the energy flows through the interactions of living and non-living components of such systems.

Economic and Social Council (ECOSOC) (经济与社会理事会) (jīng jì yǔ shè huì lǐ shì huì)

The United Nations organization serving as the central forum for discussing international economic and social issues [ECOSOC (2012)]; ECOSOC is the organ established by the Charter of the United Nations by which the State Members of the United Nations cooperate together to facilitate sustainable development.

Economic Capabilities (经济能力) (jīng jì néng lì)

The capacity of States or other actors in a socio-economic system to realize economic objectives or to function efficiently in relationship to the economic frameworks in which they are participants.

ECOSOC

See **Economic and Social Council**.

Ecosystem (生态系统) (shēng tài xì tǒng)

(1) A dynamic complex of plant, animal and micro-organism communities and their non-living environment interacting as a functional unit. [Convention on Biological Diversity, Art. 2 (1992)] (2) A dynamic complex of plant, animal, fungal, and micro-organism communities and their associated non-living environment interacting as a functional unit; the organisms living in a given environment, such as a tropical forest, a coral reef or a lake, and the physical part of the environment that impinges on them. [UNU IAS Pocket Guide, p. 13 (2007)] (3) Any natural unit or entity including living and non-living parts that interact to produce a stable system through cyclic exchange of materials. [US EPA Glossary of Climate Change Terms (2011)]

Effluent Limitation (污水限度) (wū shuǐ xiàn dù)

Any restriction imposed by the Director on quantities, discharge rates, and concentrations of pollutants which are discharged from point sources into waters of the United States, the waters of the contiguous zone, or the ocean. [US EPA Glossary of NPDES Terms (2004)]

Effluent Limitations Guidelines (ELG) (污水限度指南) (wū shuǐ xiàn dù zhǐ nán)

A regulation published by the Administrator under Section 304(b) of CWA that establishes national technology-based effluent requirements for a specific industrial category. [US EPA Glossary of NPDES Terms (2004)]

El Niño–Southern Oscillation (ENSO) (厄尔尼诺-南方波动) (è ěr ní nuò -nán fāng bō dòng)

El Niño, in its original sense, is a warm water current that periodically flows along the coast of Ecuador and Peru, disrupting the local fishery. This oceanic event is associated with a fluctuation of the intertropical surface pressure pattern and circulation in the Indian and Pacific Oceans, called the Southern Oscillation. This coupled atmosphere–ocean phenomenon is collectively known as El Niño–Southern Oscillation. During an El Niño event, the prevailing trade winds weaken and the equatorial countercurrent strengthens, causing warm surface waters in the Indonesian area to flow eastward to overlie the cold waters of the Peru current. This event has great impact on the wind, sea surface temperature, and precipitation patterns in the tropical Pacific. It has climatic effects throughout the Pacific region and in many other parts of the world. The opposite of an El Niño event is called La Niña. [US EPA Glossary of Climate Change Terms (2011)]

Elementary Flow (基本流) (jī běn liú)

Any flow of raw material entering the system being studied and which has been drawn from the environment, that is without previous human transformation; any flow of material leaving the system being studied, and which is discarded into the environment, that is without subsequent human transformation. [Von Zharen, ISO 14000, p. 196 (1996)]

Emergency Action (应急行动) (yìng jí xíng dòng)

An action performed to mitigate the impact of an emergency on human health and safety, property, or the environment. [IAEA Safety Glossary, p. 67 (2007)]

Emergency Classification (应急状态分级) (yìng jí zhuàng tài fēn jí)

The process whereby an authorized official classifies an emergency in order to declare the applicable emergency class. Upon declaration of the emergency class, the response organizations initiate the predefined response actions for that emergency class. [IAEA Safety Glossary, p. 67 (2007)]

Emergency Planning and Community Right-to-Know Act (EPCRA, United States) (应急规划和社区知情权法 (美国)) (yìng jí guī huà hé shè qū zhī qíng quán fǎ (měi guó))

The object of EPCRA, enacted in 1986, is to: (1) allow state and local planning for chemical emergencies; (2) provide for notification of emergency releases of chemicals; and (3) address communities' right-to-know about toxic and hazardous chemicals. [US EPA, EPCRA (2012)]

Eminent Domain (征用权) (zhēng yòng quán)

The government's right to take title to private property for public use upon the payment of just compensation to the landowner. [Nolon, Well Grounded, p. 448 (ELI 2001)]

Emission Standards (排污标准) (pái wū biāo zhǔn)

Rules and regulations that set limits on how much pollution can be emitted from a given source. Vehicle and equipment manufacturers have responded to many mobile source emission standards by redesigning vehicles and engines to reduce pollution. [US EPA Glossary of Mobile Source Emissions Terms (2012)]

Emissions (排放物) (pái fàng wù)

(1) The release of a substance (usually a gas when referring to the subject of climate change) into the atmosphere. [US EPA Glossary of Climate Change Terms (2011)] (2) Releases of pollutants into the air from a source, such as a motor vehicle or a factory. [US EPA Glossary of Mobile Source Emissions Terms (2012)]

Emissions Factor (排放因子) (pái fàng yīn zǐ)

A unique value for scaling emissions to activity data in terms of a standard rate of emissions per unit of activity (e.g., grams of CO_2 emitted per barrel of fossil fuel consumed). [US EPA Glossary of Climate Change Terms (2011)]

Emissions Trading/Emission Trading (排污交易) (pái wū jiāo yì)

A market established by law and regulation whereby limitations are established to cap emissions of a discharge into the atmosphere of a substance, such as sulfur dioxide, and those who reduce their emissions more than is required may trade or sell their extra reductions to others who have failed to meet their own emission reduction requirements and who can satisfy their requirements by acquiring the extra amounts to satisfy their reduction obligations. Compare **Carbon Trading**.

Employer (雇主) (gù zhǔ)

A legal person with recognized responsibility, commitment, and duties towards a worker in his or her employment by virtue of a mutually agreed relationship. (From Ref. [1].) A self-employed person is regarded as being both an employer and a worker. [IAEA Safety Glossary (2007)]

EMS

See **Environmental Management System**.

EMS Audit (环境管理体系审核) (huán jìng guǎn lǐ tǐ xì shěn hé)

A systematic and documented verification process to objectively obtain and evaluate evidence to determine whether an organization's environmental management system conforms to the EMS audit criteria set by the organization, and to communicate the results of this process to management. [Von Zharen, ISO 14000, p. 198 (1996)]

EMS Audit Criteria (环境管理体系审核标准) (huán jìng guǎn lǐ tǐ xì shěn hé biāo zhǔn)

Requirements derived from policies, practices, procedures, and other elements, as covered by ISO 14001 and, if applicable, any additional EMS requirements against which the auditor compares collected evidence about the subject matter. [Von Zharen, ISO 14000, p. 198 (1996)]

EN 45000 (EN 45000标准) (EN 45000 biāo zhǔn)

A series of standards set up by the EC to regulate and harmonize certification, accreditation, and testing activities. [Von Zharen, ISO 14000, p. 196 (1996)]

Enabling Act (授权法) (shòu quán fǎ)

Legislation passed by the New York State Legislature authorizing counties, cities, towns, and villages to carry out functions in the public interest. The power to adopt comprehensive plans, zoning ordinances, and land use regulations is delegated to towns, villages, and cities under the Town Law, Village Law, General City Law, and Municipal Home Rule Law. [Nolon, Well Grounded, p. 448 (ELI 2001)]

Enclosed or Semi-Enclosed Sea (封闭海或半封闭海) (fēng bì hǎi huò bàn fēng bì hǎi)

A gulf, basin, or sea surrounded by two or more States and connected to another sea or the ocean by a narrow outlet or consisting entirely or primarily of the territorial seas and exclusive economic zones of two or more coastal States. [Law of the Sea Convention, Art. 122 (1982)]

End State (终态) (zhōng tài)

(1) The state of radioactive waste in the final stage of radioactive waste management, in which the waste is passively safe and does not depend on institutional control. In the context of radioactive waste management, the end state includes both disposal and, if an adequate safety case can be made, indefinite storage. (2) A predetermined criterion defining the point at which a specific task or process is to be considered completed. Used in relation to decommissioning activities as the final state of decommissioning. [IAEA Safety Glossary, pp. 71–2 (2007)]

Endangered Species Act (ESA, United States) (濒危物种法 (ESA, 美国)) (bīn wēi wù zhǒng fǎ (měi guó))

[A federal statute that] protects the natural habitats of endangered and threatened species by prohibiting their import, export, transportation, and sale. A knowing violation of the act is punishable by up to . . . [1] year incarceration. [UNEP Judicial Handbook, p. 60 (2005)] Enacted in 1973 to protect threatened and endangered species.

Energy (能源) (néng yuán)

(1) Usable power such as heat or electricity. (2) The resources for producing power such as coal, oil, natural gas, biomass energy, and other kinds of resources that can be used through processing and transformation.

Energy Efficiency Label System (能源效率标识制度) (néng yuán xiào lǜ biāo shí zhì dù)

A process established by law or regulation whereby the efficiency of the use of energy by a manufactured product, machine, or vehicle is measured and required to be reported in writing on a paper attached to the product, machine, or vehicle tested, at the point when a potential purchaser would be comparing different similar products, machines, or vehicles, so that the consumer could make an informed decision about how much energy would be consumed over the expected period of use of the produce, machine, or vehicle.

Energy Intensity (能源强度) (néng yuán qiáng dù)

The ratio of energy consumption to a measure of the demand for services (e.g., number of buildings, total floorspace, floorspace-hours, number of employees, or constant dollar value of Gross Domestic Product for services). [US EPA Glossary of Climate Change Terms (2011)]

Energy Price (能源价格) (néng yuán jià gé)

The amount of money required for the sale and purchase of a specific kind of energy, such as electricity, gas, oil, or other.

Energy Saving Evaluation System
(节能考核评价制度) (jié néng kǎo hé píng jià zhì dù)

A technological measuring process to evaluate the volume of energy used, reporting both the methods used for the measurement as well as the results obtained in an open and objective manner.

Energy Saving Product Certification (节能产品认证) (jié néng chǎn pǐn rèn zhèng)

A process established by law or regulation whereby the efficiency of the use of energy, by a manufactured product, machine or vehicle, is measured and required to be reported in writing by a governmental authority or by an independent laboratory such as those associated with the International Standards Organization. See also **Energy Efficiency Label System**.

Energy Security (能源安全) (néng yúan ān quán)

The policy and practice whereby a State has sufficient sources of energy to meet the needs of its economy, without being excessively dependent on supplies that could be interrupted by actions over which the State has little to no control.

Energy Statistical Review System (能源统计审查制度) (néng yuán tǒng jì shěn chá zhì dù)

The measuring system whereby all uses of energy for given sectors of the economy can be recorded and the costs of using the energy reported, often together with any unregulated costs external to the narrow use of the energy.

Energy-saving (节约能源) (jiē yuē néng yuán)

Effective and rational utilization of energy by strengthening the management system of energy consumption, adopting flexible technologies, and enforcing reasonable economic measures which can be afforded by environment and society. Aims to reduce excessive use, mitigate energy loss, and reduce pollutants from every link of the energy production chain from source to consumption.

Enforcement (强制执行) (qiáng zhì zhí xíng)

The application by a regulatory body of sanctions against an operator, intended to correct and, as appropriate, penalize non-compliance with conditions of an authorization. [IAEA Safety Glossary, p. 72 (2007)]

Enhanced Greenhouse Effect (增强温室效应) (zēng qiáng wēn shì xiào yìng)

The concept that the natural greenhouse effect has been enhanced by anthropogenic emissions of greenhouse gases. Increased concentrations of carbon dioxide, methane, and nitrous oxide, chlorofluorocarbons (CFCs), hydrochlorofluorocarbons (HFCs), perfluorocarbons (PFCs), sulfur hexafluoride (SF_6), nitrogen trifluoride (NF_3), and other photo-chemically important gases caused by human activities such as fossil fuel consumption, trap more infra-red radiation, thereby exerting a warming influence on the climate. [US EPA Glossary of Climate Change Terms (2011)]

Environment (环境) (huán jìng)

(1) A complex of natural and anthropogenic factors and elements that are mutually interrelated and affect the ecological equilibrium and the quality of life, human health, the cultural and historical heritage, and the land-scape; that part of nature which is or could be influenced by human activity; includes natural resources both biotic and abiotic, such as air, water, soil, fauna and flora and the interactions between the same factors; property which forms part of the cultural heritage; the characteristic aspects of landscape. [UNEP Judicial Handbook, p. 4 (2005)] (2) Surroundings in which an organization operates, including air, water, land, natural resources, flora, fauna, humans, and their interrelation. The environment in this context extends from within an organization to the global system. [Von Zharen, ISO 14000, p. 196 (1996)] (3) The environment is defined broadly under the New York State Environmental Quality Review Act to include the physical conditions that will be affected by a proposed action, including land, air, water, minerals, flora, fauna, noise, resources

of agricultural, archeological, historic or aesthetic significance, existing patterns of population concentration, distribution or growth, existing community or neighborhood character, and human health. [Nolon, Well Grounded, p. 448 (ELI 2001)]

Environmental Agents (环境媒介) (huán jìng méi jiè)

Conditions other than indoor air contaminants that cause stress, comfort, and/or health problems (e.g., humidity extremes, drafts, lack of air circulation, noise, and over-crowding). [US EPA Glossary of Indoor Air Quality Terms (2011)]

Environmental Aspects (环境侧面) (huán jìng cè miàn)

Elements of an organization's activities, products, and services which can interact with the environment. [Von Zharen, ISO 14000, p. 197 (1996)]

Environmental Assessment Form (EAF) (环境评价表格) (huán jìng píng jià biǎo gé)

As used in the New York State Environmental Quality Review Act process, this is a form completed by an applicant to assist an agency in determining the environmental significance of a proposed action. A properly completed EAF must contain enough information to describe the proposed action and its location, purpose, and potential impacts on the environment. [Nolon, Well Grounded, pp. 448–9 (ELI 2001)]

Environmental Audit (环境审核) (huán jìng shěn hé)

Systematic, documented verification process of objectively obtaining and evaluating evidence to determine whether specified environmental activities, events, conditions, management systems, or information about these matters conform with audit criteria, and communicating the results of this process to the client. [Von Zharen, ISO 14000, p. 197 (1996)]

Environmental Impact (环境影响) (huán jìng yǐng xiǎng)

Any change to the environment, whether adverse or beneficial, wholly or partially resulting from an organization's activities, products, or services; consequences for human health, for the well-being of flora and fauna or for the future availability of natural resources, attributable to the input and output streams of a system. [Von Zharen, ISO 14000, p. 197 (1996)]

Environmental Impact Statement (EIS) (环境影响报告书) (huán jìng yǐng xiǎng bào gào shū)

(1) As set forth in the US National Environmental Policy Act of 1969 (NEPA), as amended: all agencies of the Federal Government shall include in every recommendation or report on proposals for legislation and other major Federal actions significantly affecting the quality of the human environment, a detailed statement by the responsible official on (i) the environmental impact of the proposed action, (ii) any adverse environmental effects which cannot be avoided should the proposal be implemented, (iii) alternatives to the proposed action, (iv) the relationship between local short-term uses of man's environment and the maintenance and enhancement of long-term productivity, and (v) any irreversible and irretrievable commitments of resources which would be involved in the proposed action should it be implemented. [NEPA Sec. 102, 42 U.S.C. § 4332]. NEPA also requires the agency to consult with other agencies with specialized expertise in the potential impacts before making the EIS. (2) A written draft or final document prepared in accordance with the New York State Environmental Quality Review Act. An EIS provides means for agencies, project sponsors, and the public to systematically consider significant, adverse environmental impact, alternatives, and mitigation strategies. An EIS facilitates the weighing of social, economic, and environmental factors in the planning and decision-making process. A Draft EIS (DEIS) is the initial statement prepared by either the project sponsor or the lead agency and circulated for review and comment before a Final EIS (FEIS) is prepared. [Nolon, Well Grounded, p. 449 (ELI 2001)]

Environmental Justice (环境正义) (huán jìng zhèng yì)

(1) The fair treatment and meaningful involvement of all people regardless of race, color, national origin, or income with respect to the development, implementation, and enforcement of environmental laws, regulations, and

policies. [US EPA Envtl. Justice (2012)] (2) Laws and procedures to ensure that all sectors of society share any burdens of environmental protection fairly, ensuring that the siting of pollution treatment facilities or other environmental protection infrastructure is not located exclusively in areas where persons live who are economically disadvantaged or of an ethnic or racial minority.

Environmental Labeling-Type I (环境标识-类型I) (huán jìng biāo shí-lèi xíng I)

Multiple criteria-based, third-party voluntary environmental labeling program. [Von Zharen, ISO 14000, p. 197 (1996)]

Environmental Management (环境管理) (huán jìng guǎn lǐ)

Parts of the overall management function of an organization that develop, implement, achieve, review, and maintain the environmental policy. [Von Zharen, ISO 14000, p. 198 (1996)]

Environmental Management System (EMS) (环境管理体系) (huán jìng guǎn lǐ tǐ xì)

(1) Organizational structure, responsibilities, practices, procedures, processes, and resources for implementing and maintaining environmental management. [Von Zharen, ISO 14000, p. 198 (1996)]; (2) A set of processes and practices that enable an organization to reduce its environmental impacts and increase its operating efficiency. [EPA Pollution Prevention http://www.epa.gov/ems/]

Environmental Monitoring Requirements (环境监测要求) (huán jìng jiān cè yāo qiú)

The duties established by law or regulations to oblige persons, government agencies, economic enterprises, or others to collect, record, and report specified data about how their conduct may affect ambient environmental conditions.

Environmental Non-Governmental Organizations (ENGOs) (非政府环境组织) (fēi zhèng fǔ huán jìng zǔ zhī)

(1) Non-governmental organizations engaged in environmental education, nature conservation, species protection, policy advocacy and many other activities. [The China Quarterly (2012)] (2) Public organizations whose primary purpose is to protect some aspect of the quality of the natural environmental and facilitate public participation in doing so.

Environmental Objectives (环境目标) (huán jìng mù biāo)

Overall environmental goals, arising from the environmental policy and significant aspects that an organization sets itself to achieve, and which are quantified wherever practicable. [Von Zharen, ISO 14000, pp. 198–9 (1996)]

Environmental Performance (环境绩效) (huán jìng jì xiào)

The measurable outputs of the environmental management system, relating to an organization's control of the impact of its activities, products, or services on the environment, based on its environmental policy, objectives, and targets. [Von Zharen, ISO 14000, p. 199 (1996)]

Environmental Policy (环境政策) (huán jìng zhèng cè)

Statement by the organization of its intentions and principles in relation to its overall environmental performance, which provides a framework for action and for the setting of its environmental objectives and targets. [Von Zharen, ISO 14000, p. 199 (1996)]

Environmental Preservation (环境保存) (huán jìng bǎo cún)

Policies and procedures aimed at conserving natural resources, preserving the natural environment, and reversing its degradation.

Environmental Quality Review (环境质量审查) (huán jìng zhí liàng shěn chá)

The process that reviewing boards must conduct to determine whether proposed projects may have significant, adverse impacts on the environment and, if they do, to study the impacts and identify alternatives and mitigation conditions that protect the environment to the maximum extent possible. [Nolon, Well Grounded, p. 449 (ELI 2001)]

Environmental Review (环境审查) (huán jìng shěn chá)

The New York State Environmental Quality Review Act [sometimes referred to as SEQRA] requires local agencies that review applications for land use approvals to take a hard look at the environmental impacts of proposed projects. Where the proposed project may have a significant adverse impact on the environment, the agency must prepare an environmental impact statement before approving the project. The adoption of comprehensive plans, zoning amendments, and other land use regulations is also subject to environmental review. [Nolon, Well Grounded, p. 449 (ELI 2001)]

Environmental Targets (环境指标) (huán jìng zhǐ biāo)

Detailed performance requirements, quantified wherever practicable, applicable to the organization or parts thereof, that arise from the environmental objectives and that need to be set and met in order to achieve those objectives. [Von Zharen, ISO 14000, p. 199 (1996)]

Environmental Tobacco Smoke (ETS) (环境性吸烟) (huán jìng xìng xī yān)

Mixture of smoke from the burning end of a cigarette, pipe, or cigar and smoke exhaled by the smoker (also secondhand smoke (SHS) or passive smoking). [US EPA Glossary of Indoor Air Quality Terms (2011)]

Environmental Treaties (环境条约) (huán jìng tiáo yuē)

Agreements between two or more States with respect to their obligations for use or protection of the environment. See **Multilateral Environmental Agreements**.

Environmentally Sound Management of Hazardous Wastes or Other Wastes (危险废物或其他废物的环境无害管理) (weí xiǎn feì wù huò qí tā feì wù de huán jìng wú haì guǎn lǐ)

Taking all practicable steps to ensure that hazardous wastes or other wastes are managed in a manner which will protect human health and the environment against the adverse effects which may result from such wastes. [Basel Convention, Art. 2, No.8 (1989)]

EPA

The United States Environmental Protection Agency.

Ergonomics (人类工程学) (rén lèi gōng chéng xué)

Applied science that investigates the impact of people's physical environment on their health and comfort (e.g., determining the proper chair height for computer operators). [US EPA Glossary of Indoor Air Quality Terms (2011)]

Erosion (侵蚀) (qīn shí)

The wearing away of land surface by water, wind, ice, or other geologic agents and by such processes as gravitational creep. [Urban Soil Primer, p. 71 (2005)]

EU (欧洲联盟) (ōu zhōu lián méng)

See **European Union**.

European Council; The Council of Europe; Europarat (欧洲理事会) (ōu zhōu lǐ shì huì)

A council of the heads of state or government of every EU country that sets the EU's general political direction and priorities and deal with complex or sensitive issues that cannot be resolved at a lower level of inter-governmental cooperation. [European Council (2012)]

European Parliament and Council Directive 94/62/Ec Of 20 December 1994 on Packaging and Packaging Waste (关于包装和包装废弃物法的欧洲议会和委员会指令 (1994年12月20日)) (guān yú bāo zhuāng hé bāo zhuāng fèi qì wù fǎ de ōu zhōu yì huì hé wéi yuán huì zhǐ lìng (1994, 12, 20))

Legal requirements for manufacturing enterprises to minimize the use of containers and avoid any unnecessary or excessive packing to reduce the volume of solid waste associated with packaging.

European Union (EU) (欧洲联盟) (ōu zhōu lián méng)

The international economic integration organization of States situated in Europe, established by treaties among the several European States.

Eutrophication (富营养化) (fù yíng yǎng huà)

The process by which water becomes enriched with plant nutrients, most commonly phosphorus and nitrogen. [US Geological Survey Circular 1206 (1995)]

Evacuation (撤离) (chè lí)

The rapid, temporary removal of people from an area to avoid or reduce short-term radiation exposure in an emergency. Evacuation is an urgent protective action (a form of intervention). If people are removed from the area for a longer period of time (more than a few months), the term relocation is used. Evacuation may be performed as a precautionary action based on plant conditions within the precautionary action zone. [IAEA Safety Glossary, p. 73 (2007)]

Evaporation (蒸发) (zhēng fā)

The process by which a substance is converted from a liquid into a vapor. "Evaporative emissions" occur when a liquid fuel evaporates and fuel molecules escape into the atmosphere. A considerable amount of hydro-carbon pollution results from evaporative emissions that occur when gasoline leaks or spills, or when gasoline gets hot and evaporates from the fuel tank or engine. [US EPA Glossary of Mobile Source Emissions Terms (2012)]

Evapotranspiration (蒸散) (zhēng sàn)

The combined process of evaporation from the Earth's surface and tran-spiration from vegetation. [US EPA Glossary of Climate Change Terms (2011)]

Evidence (证据) (zhèng jù)

Verifiable information, records or statements of fact. [Von Zharen, ISO 14000, p. 199 (1996)]

Ex-situ Conservation (迁地保护) (qiān dì bǎo hù)

(1) The conservation of components of biological diversity outside their natural habitats. [Convention on Biological Diversity, Art. 2 (1992); UNU IAS Pocket Guide, p. 13 (2007)] (2) The measures taken to preserve animals and plants in locations and sites outside of their natural or historic habitat or range (from Latin, meaning taken "from the site").

Exceptional and Emergency Situations (例外和紧急情况) (lì wài hě jǐn jí qíng kuàng)

Crises caused by natural or human-made disasters or threatened disas-ters, in which governmental authorities are required to take unusual and urgent actions to avert environmental or other harm and restore ambient environmental quality.

Exclusionary Zoning (排他分区) (pái tā fēn qū)

When a community fails to accommodate, through its zoning law, the provisions of affordable housing to meet proven regional housing needs, that community is said to practice exclusionary zoning. [Nolon, Well Grounded, p. 449 (ELI 2001)]

Executive Session (秘密会议) (mì mì huì yì)

A meeting, or part of a meeting, that is closed to the public because the topics to be discussed involve real estate, litigation, or sensitive personal matters. [Nolon, Well Grounded, p. 449 (ELI 2001)]

Exhaust Ventilation (排气通风) (pái qì tōng fēng)

Mechanical removal of air from a portion of a building (e.g., piece of equipment, room, or general area). [US EPA Glossary of Indoor Air Quality Terms (2011)]

Existing or Potentially Threatening Process (现有或具有潜在威胁的过程) (xiàn yǒu huò jù yǒu qián zài wēi xié de guò chéng)

A process applied to the soil or any other part of the ecosystem, which is or has the capability to threaten the ecological integrity of the soil. [Drafting Legislation for Sustainable Soils: A Guide, p. 91 (2004)]

Expected Consequences (预期后果) (yù qī hòu guǒ)

The results normally and reasonably anticipated as the result of an action before it is taken.

Exploitation (开发) (kāi fā)

In the context of natural resources, the decision and action to mine or extract a non-renewable resource, or harvest a renewable resource.

Exploration (勘探) (kān tàn)

The act or instance of investigating, studying, or analysing.

Export (出口) (chū kǒu)

The movement of a chemical from one Party to another Party, but excluding mere transit operations. [Rotterdam Convention, Art. 2 (1998)]

Exporter (出口商) (chū kǒu shāng)

Any person under the jurisdiction of the State of export who arranges for hazardous wastes or other wastes to be exported. [Basel Convention, Art. 2, No. 15 (1989)]

F

Facilitation (促进) (cù jìn)

A process of decision-making guided by a facilitator who insures that all affected groups and individuals are involved in a meaningful way and that the decisions are based on their input and made to achieve their mutual interests. Facilitators may be neutral outside third parties or community leaders trained or experienced in the process. [Nolon, Well Grounded, p. 449 (ELI 2001)]

Fact Sheet (情况说明书) (qíng kuàng shuō míng shū)

A document that must be prepared for all draft individual permits for NPDES [National Pollutant Discharge Elimination System] major dischargers, NPDES general permits, NPDES permits that contain variances, NPDES permits that contain sewage sludge land application plans and several other classes of permittees. The document summarizes the principal facts and the significant factual, legal, methodological and policy questions considered in preparing the draft permit and tells how the public may comment (40 CFR 124.8 and 124.56). Where a fact sheet is not required, a statement of basis must be prepared (40 CFR 124.7). [US EPA Glossary of NPDES Terms (2004)]

Family (家庭) (jiā tíng)

One or more persons occupying a dwelling as a single housekeeping unit. [Nolon, Well Grounded, p. 449 (ELI 2001)]

FAR (容积率) (róng jī lǜ)

See **Floor Area Ratio**.

Fauna (动物群) (dòng wù qún)

The term in Latin meaning all species of animals, whether domesticated animals or wildlife.

FDF (完全不同的因素) (wán quán bú tóng de yīn sù)

See **Fundamentally Different Factors**.

Federal Facility (联邦设施) (lián bāng shè shī)

Any buildings, installations, structures, land, public works, equipment, aircraft, vessels, and other vehicles and property, owned by, or constructed or manufactured for the purpose of leasing to, the federal government. [US EPA Glossary of NPDES Terms (2004)]

Federal Insecticide, Fungicide and Rodenticide Act (1994) (1994年联邦杀虫剂、杀真菌剂和灭鼠剂法) (1994 nián <lián bāng shā chóng jì, shā zhēn jùn jì hé miè shǔ jì fǎ>)

[A federal statute that] requires the registration of pesticides and prohibits that sale of pesticides not registered. A knowing violation of the act is punishable by up to . . . [1] year imprisonment. [UNEP Judicial Handbook, p. 81 (2005)]

Feedback Mechanisms (反馈机制) (fǎn kuì jī zhì)

Factors which increase or amplify (positive feedback) or decrease (negative feedback) the rate of a process. An example of positive climatic feedback is the ice-albedo feedback. [US EPA Glossary of Climate Change Terms (2011)]

Final Plat Approval (最终再分区批准) (zuì zhōng zài fēn qū pī zhǔn)

The approval by the authorized local reviewing body of a final subdivision drawing or plat that shows the subdivision, proposed improvements, and conditions as specified in the locality's subdivision regulations and as required by that body in its approval of the preliminary plat. [Nolon, Well Grounded, p. 449 (ELI 2001)]

Final Regulatory Action (最终管制行动) (zuì zhōng guǎn zhì xíng dòng)

An action taken by a Party, that does not require subsequent regulatory action by that Party, the purpose of which is to ban or severely restrict a chemical. [Rotterdam Convention, Art. 2, § (e) (1998)]

Financial Mechanism (资金机制) (zī jīn jī zhì)

A mechanism for the provision of financial resources on a grant or concessional basis, including for the transfer of technology . . . It shall function under the guidance of and be accountable to the Conference of the Parties, which shall decide on its policies, programme priorities and eligibility criteria related to this Convention. Its operation shall be entrusted to one or more existing international entities. [United Nations Framework Convention on Climate Change, Art. 11 (1992)]

Finding (认定) (rèn dìng)

A conclusion of importance based on observation(s). [Von Zharen, ISO 14000, p. 199 (1996)]

Fisheries Jurisdiction Cases (ICJ, 1974) (渔业管辖权案 (国际法院, 1974)) (yú yè guǎn xiá quán àn (guó jì fǎ yuàn, 1974))

The 1974 cases against Iceland before the International Court of Justice brought by Germany (decided 1973) and by the United Kingdom and Northern Ireland (decided 1974) regarding Iceland's extension of its fisheries jurisdiction to 50 nautical miles.

Fissile (adjective) (易裂变的 (形容词)) (yì liè biàn de (xíng róng cí))

Capable of undergoing fission by interaction with slow neutrons. More restrictive than fissionable. [IAEA Safety Glossary, p. 83 (2007)]

Fissile Material (易裂变物质) (yì liè biàn wù zhì)

Uranium-233, uranium-235, plutonium-239, plutonium-241, or any combination of these radionuclides. Excepted from this definition are: natural uranium or depleted uranium which is unirradiated; natural uranium or depleted uranium which has been irradiated in thermal reactors only. As with radioactive material, this is not a scientific definition, but one designed to serve a specific regulatory purpose. [IAEA Safety Glossary, p. 83 (2007)]

Fixed Contamination (固定污染) (gù dìng wū rǎn)

Contamination other than non-fixed contamination. See **Non-fixed Contamination**.

Floating Craft (浮动航行器) (fú dòng háng xíng qì)

A ship or other vessel to carry freight or passengers designed for use on water.

Floating Zone (浮区) (fú qū)

A zoning district that is added to the zoning law but "floats" until an application is made to apply the new district to a certain parcel. Upon the approval of the application, the zoning map is amended to apply the floating district to that parcel of land. [Nolon, Well Grounded, p. 449 (ELI 2001)]

Floodplain (洪泛区) (hóng fàn qū)

The area on the sides of a stream, river, or watercourse that is subject to periodic flooding. The extent of the floodplain is dependent on soil type, topography, and water flow characteristics. [Nolon, Well Grounded, p. 449 (ELI 2001)]

Floor Area Ratio (FAR) (容积率) (róng jī lǜ)

The gross floor area of all buildings permitted on a lot divided by the area of the lot. In zoning, the permitted building floor area is calculated by multiplying the maximum FAR specified for the zoning district by the total area of the parcel. A permitted FAR of 2 would allow the construction of 80 000 square feet of floor space on 40 000 square feet of land (40 000 × 2 = 80 000). [Nolon, Well Grounded, p. 449 (ELI 2001)]

Flora (植物群) (zhí wù qún)

A term from Latin meaning all species of plants and vegetation.

Flow Hood (流罩) (liú zhào)

Device that easily measures airflow quantity, typically up to 2500 cfm. [US EPA Glossary of Indoor Air Quality Terms (2011)]

Fluorocarbons (碳氟化合物) (tàn fú huà hé wù)

Carbon-fluorine compounds that often contain other elements such as hydrogen, chlorine, or bromine. Common fluorocarbons include chlorofluorocarbons (CFCs), hydrochlorofluorocarbons (HCFCs), hydrofluorocarbons (HFCs), and perfluorocarbons (PFCs). [US EPA Glossary of Climate Change Terms (2011)]

FOIA

See **Freedom of Information Act**.

FOIL (信息自由法) (xìn xī zì yóu fǎ)

See **Freedom of Information Law**.

Follow-up Audit (跟踪审计) (gēn zōng shěn jì)

An audit whose purpose and scope are limited to verifying that corrective action has been accomplished as scheduled and to determining that the action effectively prevented recurrence. [Von Zharen, ISO 14000, pp. 199–200 (1996)]

Food and Agriculture Organization of the United Nations (FAO) (联合国粮农组织) (lián hé guó liáng nóng zǔ zhī)

A specialized agency of the United Nations working to achieve food security for all; to make sure people have regular access to enough high-quality food to lead active, healthy lives; and to raise levels of nutrition, improve agricultural productivity, better the lives of rural populations and contribute to the growth of the world economy. [FAO (2012)]

Force Majeure (不可抗力) (bú kě kàng lì)

(1) [A]n event that is unforeseeable, unavoidable, and external that makes execution impossible. Certain events, beyond the control of the parties, may inhibit the parties from fulfilling their duties and obligations under ... project agreements [or contracts]. [World Bank, Force Majeure Clauses (2012)] (2) An event caused by circumstances beyond the control of an affected facility. The event prevents the owner from conducting required performance tests on time despite best efforts to do so. [US EPA, OAR Policy Guidance Fact Sheet (2011)] (3) The term in French for an unexpected or overwhelming event that is so great a force that it makes it impossible for a party to perform an agreed duty and relieves it temporarily during the event of its obligation to act.

Forcing Mechanism (强制机制) (qiáng zhì jī zhì)

A process that alters the energy balance of the climate system, i.e., changes the relative balance between incoming solar radiation and outgoing infrared radiation from Earth. Such mechanisms include changes in solar irradiance, volcanic eruptions, and enhancement of the natural green-house effect by emissions of greenhouse gases. [US EPA Glossary of Climate Change Terms (2011)]

Forum Non Conveniens (不方便法院) (bù fāng biàn fǎ yuàn)

Usually a prudential doctrine allowing a court to dismiss a case when another more convenient forum exists. This issue is most likely to arise in the environmental context when a case concerns transboundary environmental harm. [UNEP Judicial Handbook, p. 68 (2005)]

Fossil (化石) (huà shí)

Recognizable remains such as bones, shells, leaves, burrows, impressions, or tracks of past life on the Earth. [USGS, Fossils (2012)]

Fossil Fuels (矿物燃料) (kuàng wù rán liào)

(1) Fuels – such as coal, natural gas, and crude oil – that come from the compressed remains of ancient plants and animals. Gasoline and diesel are fossil fuels that can be burned in internal combustion engines to power everything from jet planes to automobiles to railroad locomotives. [US EPA Glossary of Mobile Source Emissions Terms (2012)] (2) Coal, gas, oil, tar sands, and other substances resulting from ancient depositions of animal and plant matter that are compressed and may be extracted and refined and burned to produce electricity or serve other energy objectives.

Four-Zero-One (401) (A) CERTIFICATION (401 (A)认证) (401 (A)rèn zhèng)

A requirement of Section 401 (a) of the Clean Water Act that all federally issued permits be certified by the state in which the discharge occurs. The state certifies that the proposed permit will comply with state water quality standards and other state requirements. [US EPA Glossary of NPDES Terms (2004)]

Framework Convention on Civil Defence Assistance (民防援助框架公约) (mín fáng yuán zhù kuàng jià gōng yuē)

A treaty ratified by the United Nations May 22, 2000 (21 votes), and entered into force September 23, 2001. [Convention on Civil Defense Assistance (2000)]

Freedom of Information Act (FOIA) (信息自由法) (xìn xī zì yóu fǎ)

A federal law, originally enacted in 1966, providing a procedure under which interested persons may request, and US government agencies must disclose, if not exempted or excused, all or appropriate parts of previously unreleased official US government documents and information. FOIA has been amended several times and there are a number of exemptions to this statute. [Freedom of Information Act, 5 U.S.C. § 552]

Freedom of Information Law (信息自由法) (xìn xī zì yóu fǎ)

The New York State Freedom of Information Law [often called FOIL] requires that the public be provided access to government records, including local land use document, such as photos, maps, designs, drawings, rules, regulations, codes, manuals, reports, files, and opinions. Public access may be denied if it would constitute an invasion of privacy. [Nolon, Well Grounded, pp. 449–50 (ELI 2001)]

Freedom of the High Seas (公海自由) (gōng hǎi zì yóu)

The high seas are open to all States, whether coastal or land-locked. Freedom of the high seas is exercised under the conditions laid down by this Convention and by other rules of international law. It comprises . . . both for coastal and land-locked States: freedom of navigation; freedom of overflight; freedom to lay submarine cables and pipelines, subject to Part VI; . . . freedom to construct artificial islands and other installations permitted under international law, subject to Part VI; . . . freedom of fishing, subject to the conditions laid down in section 2; . . . freedom of scientific research, subject to Parts VI and XIII . . . These freedoms shall be exercised by all States with due regard for the interests of other States in their exercise of the freedom of the high seas, and also with due regard for the rights under this Convention with respect to activities in the Area. [Law of the Sea Convention, Art. 87 (1982)]

Freight Container (货物集装箱) (huò wù jí zhuāng xiāng)

An article of transport equipment designed to facilitate the transport of goods, either packaged or unpackaged, by one or more modes of transport without intermediate reloading, which is of a permanent enclosed character, rigid and strong enough for repeated use, and must be fitted with devices facilitating its handling, particularly in transfer between conveyances and from one mode of transport to another. A small freight container is that which has either any overall outer dimension less than 1.5 m, or an internal volume of not more than 3 m^3. Any other freight container is considered to be a large freight container. [IAEA Safety Glossary, p. 85 (2007)]

French Environmental Code (法国环境法典) (fǎ guó huán jìng fǎ diǎn)

An environmental code set out by the government of France (Act no. 2002-276 of 27 February 2002, Article 132 Official Journal of 28 February 2002). [French Environmental Code]

Freshwater Wetlands Regulations (淡水湿地规章) (dàn shuǐ shī dì guī zhāng)

Laws passed by federal, state, and local governments to protect wetlands by limiting the types and extent of activities permitted within the wetlands. These laws require land owners to secure permits before conducting many activities, such as draining, filling, or constructing buildings. [Nolon, Well Grounded, p. 450 (ELI 2001)]

Frigate Mackerel (舵鲣) (duò jiān)

Auxis thazard; Auxis rochei. A subspecies of tuna found in tropical waters. [Law of the Sea Convention, Annex I (1982)]

Frontage (正面) (zhèng miàn)

Zoning laws typically require that developable lots have specified numbers of linear feet that front on a dedicated street. A 100-foot frontage means that a lot must have 100 linear feet on the side of the parcel that fronts [faces] on a street. [Nolon, Well Grounded, p. 450 (ELI 2001)]

Fukushima Nuclear Power Plant Accident in Japan (日本福岛核电站核泄漏事故) (rì běn fú dǎo hé diàn zhàn hé xiè lòu shì gù)

A nuclear accident in the Tohoku district of Japan following an earthquake and tsunami on March 11, 2011. [Report of the Japanese Government, I-1 (2011)]

Full and Open Exchange of Information (充分和开放的信息交流) (chōng fèn hé kāi fàng dē xìn xī jiāo liú)

(1) A principle endorsed by OECD and other international organizations to facilitate the exchange of scientific information among nations in order to enable collaborative environmental problem-solving. [National Academies Press] (2) The provision of data or other information on the basis of a law, regulation or agreement to disclose the data to the public, usually to ensure effective public participation in environmental decision-making.

Functional Unit (功能单元) (gōng néng dān yuán)

A measure of performance of the main functional outputs of the product or service system. [Von Zharen, ISO 14000, p. 200 (1996)]

Fundamentally Different Factors (FDF) (完全不同的因素) (wán quán bú tóng de yīn sù)

Those components of a petitioner's facility that are determined to be so unlike those components considered by EPA during the effluent limitation guideline and pretreatment standards rulemaking that the facility is worthy of a variance from the effluent limitations guidelines or categorical pretreatment standards. [US EPA Glossary of NPDES Terms (2004)]

Fungi (菌类) (jùn lèi)

Any of a group of parasitic lower plants that lack chlorophyll, including molds and mildews. [US EPA Glossary of Indoor Air Quality Terms (2011)]

G

Gabcikovo–Nagymaros Dam Case (Hungary and Slovakia) (ICJ, 1997) (加布奇科沃-大毛罗斯大坝案) (匈牙利与斯洛伐克 (国际法院, 1997年)) (jiā bù qí kē wò -dà máo luó sī dà bà àn (xiōng yá lì yǔ sī luò fá kè) (guó jì fǎ yuàn, 1997 nián))

A 1997 decision and an ongoing case before the International Court of Justice (ICJ) regarding a dam on the Danube between Hungary and Slovakia designed to prevent floods and create electricity. Hungary abandoned its work on the project citing environmental concerns, and Slovakia brought the case to the ICJ. After a decision largely in Slovakia's favor, political changes caused the case to be reopened and it is still pending. [Gabcikovo–Nagymaros Project (1997)]

Gas Sorption (气体吸附) (qì tǐ xī fù)

Devices used to reduce levels of airborne gaseous compounds by passing the air through materials that extract the gases. The performance of solid sorbents is dependent on the airflow rate, concentration of the pollutants, presence of other gases or vapors, and other factors. [US EPA Glossary of Indoor Air Quality Terms (2011)]

GCM (大气环流模型) (dà qì huán liú mó xíng)

See **General Circulation Model**.

GEF (全球环境基金) (quán qiú huán jìng jī jīn)

See **Global Environment Facility**.

General Assembly (联合国大会) (lián hé guó dà huì)

(1) An organ of the United Nations established by the Charter of the United Nations, as the principal assembly of State Members of the United

Nations to facilitate their cooperation and collective decision-making under the Charter. (2) [T]he main deliberative, policymaking and representative organ of the United Nations. [Comprised of all] Members of the United Nations, it provides a unique forum for multilateral discussion of the full spectrum of international issues covered by the [United Nations] Charter. The Assembly meets in regular session intensively from September to December each year, and thereafter as required. [United Nations, About the General Assembly (2012)]

General Circulation Model (GCM) (大气环流模型) (dà qì huán liú mó xíng)

A global, three-dimensional computer model of the climate system which can be used to simulate human-induced climate change. GCMs are highly complex and they represent the effects of such factors as reflective and absorptive properties of atmospheric water vapor, greenhouse gas concentrations, clouds, annual and daily solar heating, ocean temperatures and ice boundaries. The most recent GCMs include global representations of the atmosphere, oceans, and land surface. [US EPA Glossary of Climate Change Terms (2011)]

General Permit (一般许可证) (yī bān xǔ kě zhèng)

A NPDES permit issued under 40 CFR 122.28 that authorizes a category of discharges under the CWA within a geographical area. A general permit is not specifically tailored for an individual discharger. [US EPA Glossary of NPDES Terms (2004)]

Generator (产生者) (chǎn shēng zhě)

Any person whose activity produces hazardous wastes or other wastes or, if that person is not known, the person who is in possession and/or control of those wastes. [Basel Convention, Art. II, § 18 (1989)]

Genetic Material (遗传物质) (yí chuán wù zhì)

Any material of plant, animal, microbial, or other origin containing functional units of heredity. [UNU IAS Pocket Guide, p. 13 (2007)]

Genetic Resources (遗传资源) (yí chuán zī yuán)

Genetic material of actual or potential value. [Convention on Biological Diversity Art. 2 (1992); UNU IAS Pocket Guide, p. 13 (2007)]

Genus ([生物]属) ([shēng wù] shǔ)

A way of categorizing organisms into groups with similar character-istics that is used in taxonomy to name animal and plant species. The first (Latin) part of the formalized, scientific name given to a species. For example, the gray bat is part of the genus *Myotis* and the scientific name for its species is *Myotis grisenscens*. [US Fish & Wildlife Service, Endangered Species Glossary (2012)]

Geographically Disadvantaged States (地理不利国) (dì lǐ bú lì guó)

Coastal States, including States bordering enclosed or semi-enclosed seas, whose geographical situation makes them dependent upon the exploita-tion of the living resources of the exclusive economic zones of other States in the subregion or region for adequate supplies of fish for the nutritional purposes of their populations or parts thereof, and coastal States which can claim no exclusive economic zones of their own. [Law of the Sea Convention, Art. 70 (1982)]

Geosphere (岩石圈) (yán shí quān)

(1) The soils, sediments, and rock layers of the Earth's crust, both con-tinental and beneath the ocean floors. [US EPA Glossary of Climate Change Terms (2011)] (2) Those parts of the lithosphere not considered to be part of the biosphere. In safety assessment, usually used to distinguish the subsoil and rock (below the depth affected by normal human activities, in particular agriculture) from the soil that is part of the biosphere. [IAEA Safety Glossary, p. 87 (2007)]

Geothermal Energy (地热能源) (dì rè néng yuán)

(1) Heat energy from the interior earth that can be used for power/electricity or heat. It is considered a renewable resource. [US Department of Energy, Geothermal Energy & the Environment (2007)] (2) Heat or steam derived from the molten core of the Earth and used to produce electricity or to heat buildings or for industrial and other uses.

German Environmental Law (德国环境法) (dé guó huán jìng fǎ)

The environmental law of Germany, currently a complicated collection of unconnected laws. Germany has attempted for a number of years to create a comprehensive single Environmental Code, and has a draft comprising five books covering different areas of the law. However, as of 2012 the draft Code has yet to be passed into law. [Germany Federal Environment Agency]

GHG (温室气体) (wēn shì qì tǐ)

See **Greenhouse Gas**.

Glacier (冰川) (bīng chuān)

A multi-year surplus accumulation of snowfall in excess of snowmelt on land and resulting in a mass of ice at least 0.1 km^2 in area that shows some evidence of movement in response to gravity. A glacier may terminate on land or in water. Glacier ice is the largest reservoir of fresh water on Earth and is second only to the oceans as the largest reservoir of total water. Glaciers are found on every continent except Australia. [US EPA Glossary of Climate Change Terms (2011)]

Global Crop Diversity Trust (全球农作物多样性信托基金) (quán qiú nóng zuò wù duō yàng xìng xìn tuō jī jīn)

The Global Crop Diversity Trust was founded by the United Nations Food and Agriculture Organisation (FAO) and Bioversity International, acting on behalf of the foremost international research organizations in this field (CGIAR). The Trust is currently hosted in Rome by FAO. [Global Crop Diversity Trust, Founders (2012)] The trust is a unique public–private partnership raising funds from individual, corporate and government donors to establish an endowment fund that will provide complete and continuous funding for key crop collections . . . [Its] goal is to advance an efficient and sustainable global system of . . . conservation by promoting the rescue, understanding, use and long-term conservation of valuable plant genetic resources. [Global Crop Diversity Trust, About Us (2012)]

Global Environment Facility (GEF) (全球环境基金) (quán qiú huán jìng jī jīn)

An independent financial organization, uniting 182 member governments, that provides grants to developing countries and countries with economies in transition for projects related to biodiversity, climate change, international waters, land degradation, the ozone layer, and persistent organic pollutants. [GEF, About GEF (2011)]

Global Environment Facility (GEF) Council (全球环境基金理事会) (quán qiú huán jìng jī jīn lǐ shì huì)

The main governing body of the GEF. It functions as an independent board of directors, with primary responsibility for developing, adopting, and evaluating GEF programs. [GEF, GEF Council (2012)]

Global Environment Facility (GEF) Eligibility Criteria (全球环境基金的资格标准) (quán qiú huán jìng jī jīn de zī gé biāo zhǔn)

[C]ountries are eligible for GEF funding in a focal area if: [t]hey meet eligibility criteria established by the relevant COP of that convention; [t]hey are members of the conventions and are countries eligible to borrow from the World Bank (IBRD and/or IDA); [t]hey are eligible recipients of UNDP technical assistance through country programming. [GEF, Country Eligibility (2012)]

Global Environment Facility (GEF) Trust Fund
(全球环境基金信托基金) (quán qiú huán jìng jī jīn xìn tuō jī jīn)

The GEF administers three trust funds, the Global Environment Facility Trust Fund (GEF); Least Developed Countries Trust Fund (LDCF); Special Climate Change Trust Fund (SCCF); the Nagoya Protocol Implementation Fund (NPIF) and provides secretariat services, on an interim basis, for the Adaptation Fund. [GEF, GEF-Administered Trust Funds (2012)]

Global Warming (全球变暖) (quán qiú biàn nuǎn)

Global warming is an average increase in the temperature of the atmosphere near the Earth's surface and in the troposphere, which can contribute to changes in global climate patterns. Global warming can occur from a variety of causes, both natural and human induced. In common usage, "global warming" often refers to the warming that can occur as a result of increased emissions of greenhouse gases from human activities. [US EPA Glossary of Climate Change Terms (2011)]

Global Warming Potential (GWP) (全球变暖潜势) (quán qiú biàn nuǎn qián shì)

[T]he cumulative radiative forcing effects of a gas over a specified time horizon resulting from the emission of a unit mass of gas relative to a reference gas. The GWP-weighted emissions of direct greenhouse gases in the US Inventory are presented in terms of equivalent emissions of carbon dioxide (CO_2), using units of teragrams of carbon dioxide equivalents (Tg CO_2 Eq.). Conversion: Tg = 109 kg = 106 metric tons = 1 million metric tons. The molecular weight of carbon is 12, and the molecular weight of oxygen is 16; therefore, the molecular weight of CO_2 is 44 (i.e., 12 + [16 × 2]), as compared to 12 for carbon alone. Thus, carbon comprises 12/44 ths of carbon dioxide by weight. [US EPA Glossary of Climate Change Terms (2011)]

Grab Sample (随机采取的样品) (suí jī cǎi qǔ de yàng pǐn)

A sample which is taken from a wastestream on a one-time basis without consideration of the flow rate of the wastestream and without consideration of time. [US EPA Glossary of NPDES Terms (2004)]

Grain Size (结晶尺寸) (jié jīng chǐ cùn)

Refers to the diameter of individual grains of sediment or the lithified particles in clastic rocks or other granular materials. [C.K. Wentworth, *Clastic Sediments* (1922)]

Gravel (砾石) (lì shí)

Rounded or angular fragments of rock as much as 3 inches (2 mm to 7.6 cm) in diameter. [Urban Soil Primer 72 (2005)]

Great Lakes Water Quality Agreement (United States and Canada) (五大湖水质协定 (美国加拿大)) (wǔ dà hú shuǐ zhì xié dìng (měi guó jiā ná dà))

A 1978 agreement between the US and Canada regarding each country's commitment to restore and maintain the chemical, physical and biological integrity of the Great Lakes Basin Ecosystem. [Int'l Joint Comm'n (2012)]

Green Economy (绿色经济) (lǜ sè jīng jì)

An economy that results in improved human well-being and social equity, while significantly reducing environmental risks and ecological scarcities. [UNEP Green Economy Report, p. 16 (2011)]

Greenhouse Effect (温室效应) (wēn shì xiào yìng)

Trapping and building-up of heat in the atmosphere (troposphere) near the Earth's surface. Some of the heat flowing back toward space from the Earth's surface is absorbed by water vapor, carbon dioxide, ozone, and several other gases in the atmosphere and then reradiated back toward the Earth's surface. If the atmospheric concentrations of these greenhouse gases rise, the average temperature of the lower atmosphere will gradually increase. [US EPA Glossary of Climate Change Terms (2011)]

Greenhouse Gas (GHG) (温室气体) (wēn shì qì tǐ)

Any gas that absorbs infrared radiation in the atmosphere. Greenhouse gases include, but are not limited to, water vapor, carbon dioxide (CO_2), methane (CH_4), nitrous oxide (N_2O), chlorofluorocarbons (CFCs), hydrochlorofluorocarbons (HCFCs), ozone (O_3), hydrofluorocarbons (HFCs), perfluorocarbons (PFCs), and sulfur hexafluoride (SF_6). [US EPA Glossary of Climate Change Terms (2011)]

Greenpeace (绿色和平) (lǜ sè hé píng)

[A]n independent [non-governmental] global campaigning organization that acts to change attitudes and behavior, to protect and conserve the environment and to promote peace. [Greenpeace, About Greenpeace (2012)]

Gulf Oil Spill (墨西哥湾漏油事件) (mò xī gē wān lòu yóu shì jiàn)

A human, economic, and environmental disaster in which four million gallons of oil were released into the Gulf of Mexico as a result of an explosion on a deepwater oil drilling rig on April 20, 2010. [Deepwater Report to the President, p. vi (2011)]

GWP (全球变暖潜势) (quán qiú biàn nuǎn qián shì)

See **Global Warming Potential**.

H

Habitat (栖息地) (qī xī dì)

The place or type of site where an organism or population naturally occurs. [Convention on Biological Diversity, Art. 2 (1992)]

Habitat/Species Management Area: Protected Area Managed Mainly for Conservation through Management Intervention (栖息地/物种管理区域：通过管理性干预来实现自然保护区自然环境保护) (qī xī dì/wù zhǒng guǎn lǐ qū yù: tōng guò guǎn lǐ xìng gàn yù lái shí xiàn zì rán bǎo hù qū zì rán huán jìng bǎo hù)

Area of land and/or sea subject to active intervention for management purposes so as to ensure the maintenance of habitats and/or to meet the requirements of specific species. [Guidelines for Protected Area Management Categories, p. 19 (1994)]

Halocarbons (卤化碳) (lǔ huà tàn)

Compounds containing either chlorine, bromine or fluorine, and carbon. Such compounds can act as powerful greenhouse gases in the atmosphere. The chlorine and bromine containing halocarbons are also involved in the depletion of the ozone layer. [US EPA Glossary of Climate Change Terms (2011)]

Hard Bedrock (坚硬基石) (jiān yìng jī shí)

Bedrock that cannot be excavated except by blasting or by the use of special equipment that is not commonly used in construction. [Urban Soil Primer, p. 72 (2005)]

Harmonized Procedures (协调一致的程序) (xié tiáo yī zhì de chéng xù)

Methods and systems, agreed among different parties that adopt the same procedures and protocols, to govern how their comparable activities are to be conducted in the same manner.

Hazardous Substance (有害物质) (yǒu hài wù zhì)

Any substance, other than oil, which, when discharged in any quantities into waters of the US, presents an imminent and substantial danger to the public health or welfare, including but not limited to fish, shellfish, wildlife, shorelines, and beaches (Section 311 of the CWA); identified by EPA as the pollutants listed under 40 CFR Part 116. Any substance, other than oil, which, when discharged in any quantities into waters of the United States, presents an imminent and substantial danger to the public health or welfare, including but not limited to fish, shellfish, wildlife, shorelines, and beaches (Section 311 of the CWA); identified by EPA as the pollutants listed under 40 CFR Part 116. [US EPA Glossary of NPDES Terms (2004)]

Haze (阴霾) (yīn mái)

Atmospheric particulate matter and gases that diminish visibility. Visibility is reduced when light encounters tiny pollution particles, such as soot and dust, and some gases (such as nitrogen dioxide) in the air. Some light is absorbed by the particles and gases and other light is scattered away before it reaches your eye. More pollutants mean more absorption and scattering of light, resulting in more haze. Some haze-causing pollutants are directly emitted to the atmosphere from vehicle emissions; others are formed indirectly when pollutants from mobile sources react with other elements and materials in the atmosphere. [US EPA Glossary of Mobile Source Emissions Terms (2012)]

Heat Islands (热岛效应) (rè dǎo xiào yìng)

Small areas of artificially drained urban soils surrounded by tall buildings that change the soil temperature and moisture patterns. May also refer to an entire city with an artificial microclimate. [Urban Soil Primer, p. 72 (2005)]

Helsinki Rules

See **Use of Water Resources of International Rivers, the Helsinki Rules**.

Heritage (遗产) (yí chǎn)

Something that can be inherited or passed down from preceding generations. [UNEP Judicial Handbook, p. 109 (2005)]

Historic District (历史区域) (lì shǐ qū yù)

A regulatory overlay zone within which new developments must be compatible with the architecture of the district's historic structures. Alterations and improvements of historic structures must involve minimum interference with the historic features of the buildings. [Nolon, Well Grounded, p. 450 (ELI 2001)]

Historic Preservation Commission (历史保存委员会) (lì shǐ bǎo cún wěi yuán huì)

A commission established to review proposed projects within historic districts for compliance with standards for new development or alteration or improvement of historic buildings or landmarks. [Nolon, Well Grounded, p. 450 (ELI 2001)]

Histosols (有机土) (yǒu jī tǔ)

Organic soils that have organic soil materials in more than half of the upper 80 cm (32 inches) or that have organic materials of any thickness if they overlie rock or fragmental materials that have interstices filled with organic soil materials. [US EPA Glossary of Wetlands Terms (2011)]

Home Occupation (居家办公) (jū jiā bàn gōng)

A business conducted in a residential dwelling unit that is incidental and subordinate to the primary residential use. Regulations of home occupations usually restrict the percentage of the unit that can be used for the occupation, the exterior evidence of the business, and the amount of parking allowed and traffic generated. [Nolon, Well Grounded, p. 450 (ELI 2001)] "Home occupation" is the term for running a business from one's home in an area zoned for residences rather than one zoned for commercial or office uses. Due to the potential for increased noise and traffic in a residential neighborhood, many municipalities enforce special regulations for conducting such businesses.

Home Rule Authority (地方自治授权) (dì fāng zì zhì shòu quán)

Gives local governments the power to adopt laws relating to their local property, affairs, and governments in addition to the powers specifically delegated to them by the legislature. [Nolon, Well Grounded, p. 450 (ELI 2001)]

Human Factors Engineering (人因工程) (rén yīn gōng chéng)

Engineering in which factors that could influence human performance are taken into account. [IAEA Safety Glossary, p. 92 (2007)]

Humidifier Fever (湿热症) (shī rè zhèng)

A respiratory illness caused by exposure to toxins from microorganisms found in wet or moist areas in humidifiers and air conditioners. Also called air conditioner or ventilation fever. [US EPA Glossary of Indoor Air Quality Terms (2011)]

Humus (腐殖质) (fǔ zhí zhì)

The well-decomposed organic matter in mineral soils. [Urban Soil Primer, p. 72 (2005)]

Hydrocarbons (碳氢化合物) (tàn qīng huà hé wù)

(1) Substances containing only hydrogen and carbon. Fossil fuels are made up of hydrocarbons. [US EPA Glossary of Climate Change Terms (2011)] (2) Chemical compounds that contain hydrogen and carbon. Most motor vehicles and engines are powered by hydrocarbon-based fuels such as gasoline and diesel. Hydrocarbon pollution results when unburned or partially burned fuel is emitted from the engine as exhaust, and also when fuel evaporates directly into the atmosphere. Hydrocarbons include many toxic compounds that cause cancer and other adverse health effects. Hydrocarbons also react with nitrogen oxides in the presence of sunlight to form ozone. Hydrocarbons, which may take the form of gases, tiny particles, or droplets, come from a great variety of industrial and natural processes. In typical urban areas, a very significant fraction comes from cars, buses, trucks, and non-road mobile sources such as construction vehicles and boats. [US EPA Glossary of Mobile Source Emissions Terms (2012)]

Hydrochlorofluorocarbons (HCFCs) (氢氯氟碳化合物) (qīng lù fú tàn huà hé wù)

Compounds containing hydrogen, fluorine, chlorine, and carbon atoms. Although ozone depleting substances, they are less potent at destroying stratospheric ozone than chlorofluorocarbons (CFCs). They have been introduced as temporary replacements for CFCs and are also greenhouse gases. [US EPA Glossary of Climate Change Terms (2011)]

Hydrofluorocarbons (HFCs) (氟烃化合物) (fú tīng huà hé wù)

Compounds containing only hydrogen, fluorine, and carbon atoms. They were introduced as alternatives to ozone depleting substances in serving many industrial, commercial, and personal needs. HFCs are emitted as by-products of industrial processes and are also used in manufacturing. They do not significantly deplete the stratospheric ozone layer, but they are powerful greenhouse gases with global warming potentials ranging from 140 (HFC-152a) to 11 700 (HFC-23). [US EPA Glossary of Climate Change Terms (2011)]

Hydrologic Cycle (水循环) (shuǐ xún huán)

The process of evaporation, vertical and horizontal transport of vapor, condensation, precipitation, and the flow of water from continents to oceans. It is a major factor in determining climate through its influence on surface vegetation, the clouds, snow and ice, and soil moisture. The hydrologic cycle is responsible for 25 to 30 percent of the mid-latitudes' heat transport from the equatorial to polar regions. [US EPA Glossary of Climate Change Terms (2011)]

Hydrologic Soil Groups (水文土壤组) (shuǐ wén tǔ rǎng zǔ)

Refers to soils grouped according to their runoff potential. The soil properties that influence this potential are those that affect the minimum rate. [Urban Soil Primer, p. 72 (2005)]

Hydrologic Unit (水文单位) (shuǐ wén dān wèi)

Catchment areas with an outlet in or affecting a densely populated area. [Urban Soil Primer, p. 72 (2005)]

Hydrology (水文学) (shuǐ wén xué)

(1) The scientific study of water. Hydrology has evolved as a science in response to the need to understand the complex water systems of the Earth and help solve water problems. [USGS, Hydrology Primer (2011)] (2) The study of physical geography dealing with the distribution and flow of waters above and beneath the land, and the relationship of waters to human actions.

Hydropower (水电) (shuǐ diàn)

(1) The use of water to turn turbines and water wheels to produce electricity or other forms of energy. (2) An energy form that uses the Earth's water cycle to generate electricity. Water evaporates from the Earth's surface, forms clouds, precipitates back to earth, and flows toward the ocean. The movement of water as it flows downstream creates kinetic energy that can be converted into electricity. A hydroelectric power plant

converts this energy into electricity by forcing water, often held at a dam, through a hydraulic turbine that is connected to a generator. The water exits the turbine and is returned to a stream or riverbed below the dam. [US EPA, Hydroelectricity (2007)]

Hydrosphere (水圈) (shuǐ quān)

The component of the climate system comprising liquid surface and subterranean water, such as oceans, seas, rivers, fresh water lakes, underground water, etc. [US EPA Glossary of Climate Change Terms (2011)]

Hypersensitivity Diseases (过敏性疾病) (guò mǐn xìng jí bìng)

Diseases characterized by allergic responses to pollutants. The hypersensitivity diseases most clearly associated with indoor air quality are asthma, rhinitis, and hypersensitivity pneumonitis. Hypersensitivity pneumonitis is a rare but serious disease that involves progressive lung damage as long as there is exposure to the causative agent. [US EPA Glossary of Indoor Air Quality Terms (2011)]

Hypersensitivity Pneumonitis (过敏性肺炎) (guò mǐn xìng fèi yán)

A group of respiratory diseases that cause inflammation of the lung (specifically granulomatous cells). Most forms of hypersensitivity pneumonitis are caused by the inhalation of organic dusts, including molds. [US EPA Glossary of Indoor Air Quality Terms (2011)]

Hypoxic Zone (缺氧区) (quē yǎng qū)

An area within aquatic environments where the dissolved oxygen (DO) concentration is reduced to the point where it becomes detrimental to aquatic organisms (1–30% saturation of DO of < 2 mg/l). [NOAA Gulf of Mexico Hypoxia Assessment]

I

IAQ

See **Indoor Air Quality**.

IAQ Backgrounder (室内空气质量简报) (shì nèi kōng qì zhì liàng jiǎn bào)

A component of the IAQ Tools for Schools Action Kit that provides a general introduction to IAQ issues, as well as IAQ program implementation information. [US EPA Glossary of Indoor Air Quality Terms (2011)]

IAQ Checklist (室内空气质量检查表) (shì nèi kōng qì zhì liàng jiǎn chá biǎo)

A component of the IAQ Tools for Schools Action Kit containing information and suggested easy-to-do activities for school staff to improve or maintain good indoor air quality. Each Activity Guide focuses on topic areas and actions that are targeted to particular school staff. The Checklists are to be completed by the staff and returned to the IAQ Coordinator as a record of activities completed and assistance as requested. [US EPA Glossary of Indoor Air Quality Terms (2011)]

IAQ Coordinator (室内空气质量协调员) (shì nèi kōng qì zhì liàn xié tiáo yuán)

An individual at the school and/or school district level who provides leadership and coordination of IAQ activities. [US EPA Glossary of Indoor Air Quality Terms (2011)]

IAQ Management Plan (室内空气质量管理计划) (shì nèi kōng qì zhì liàng guǎn lǐ jì huá)

A component of the IAQ Tools for Schools Kit, specifically, a set of flexible and specific steps for preventing and resolving IAQ problems. [US EPA Glossary of Indoor Air Quality Terms (2011)]

IAQ Team (室内空气质量管理团队) (shì nèi kōng qì zhì liàng guǎn lǐ tuán duì)

People who have a direct impact on IAQ in the schools (school staff, administrators, school board members, students and parents) and who implement the IAQ Action Packets. [US EPA Glossary of Indoor Air Quality Terms (2011)]

Ice Core (冰核) (bīng hé)

A cylindrical section of ice removed from a glacier or an ice sheet in order to study climate patterns of the past. By performing chemical analyses on the air trapped in the ice, scientists can estimate the percentage of carbon dioxide and other trace gases in the atmosphere at a given time. [US EPA Glossary of Climate Change Terms (2011)]

Ice Sheet (冰原) (bīng yuán)

A glacier of considerable thickness and more than 50 000 sq. km in area. It forms a continuous cover of ice and snow over a land surface. An ice sheet is not confined by the underlying topography but spreads outward in all directions. During the Pleistocene Epoch, ice sheets covered large parts of North America and northern Europe but they are now confined to polar regions (e.g., Greenland and Antarctica). [NASA Earth Observatory (2012)]

Illegal Traffic (非法运输) (fēi fǎ yùn shū)

Any transboundary movement of hazardous wastes or other wastes as specified in Article 9 of the Basel Convention. [Basel Convention, Art. 2, No. 21 (1989)]

Illicit Trafficking (in Nuclear or Radioactive Material)
非法贩运 (核材料或放射性物质) (fēi fǎ fàn yùn (hé cái liào huò fàng shè xìng wù zhì))

(1) The term is in use but there is no agreed definition. The vague phrase is used in different contexts to mean different things. The term nuclear trafficking is vaguer still and more open to (mis)interpretation, and is better avoided where clear communication is necessary. [IAEA Safety Glossary, p. 93 (2007)] (2) In general terms, the illegal movement of nuclear or radioactive material from its official custody.

Immunity of Warships on the High Seas (公海上军舰豁免权) (gōng hǎi shàng jūn jiàn huò miǎn quán)

Warships on the high seas have complete immunity from the jurisdiction of any State other than the flag State. [Law of the Sea Convention, Art. 95 (1982)]

Impact Assessment (影响评估) (yǐng xiǎng píng gū)

Phase of Life Cycle Assessment aimed at understanding and evaluating the magnitude and significance of environmental impacts based on the life cycle inventory analysis. [Von Zharen, ISO 14000, p. 200 (1996)]

Impervious Soil (不可渗透的土壤) (bù kě shèn tòu de tǔ rǎng)

A soil through which water, air, or roots penetrate slowly or not at all. No soil is absolutely impervious to air and water all the time. [Urban Soil Primer, p. 72 (2005)]

Implementation of International Environmental Law (国际环境法的实施) (guó jì huán jìng fǎ dē shí shī)

The legislation and regulations whereby a State establishes at the national and sub-national levels the means to observe international obligations, such as those agreed in treaties and multilateral environmental agreements.

Implementation Plan or Measures (实施计划或措施) (shí shī jì huà huò cuò shī)

A plan or measures to coordinate all the related strategies that are to be carried out to achieve the objectives contained in the comprehensive plan. An implementation plan answers the questions: who, what, where, how. [Nolon, Well Grounded, p. 450 (ELI 2001)]

Import (进口) (jìn kǒu)

The movement of a chemical from one Party to another Party, but excluding mere transit operations. [Rotterdam Convention, Art. 2 (1998)]

Imported Cigarettes Case (United States v. Thailand) (BISD 37S/200, 1990) (进口香烟案 (美国诉泰国) (BISD 37S/200, 1990)) (jìn kǒu xiāng yān àn (měi guó su tài guó))

A 1990 GATT decision regarding the restrictions on importation of and internal taxes on cigarettes maintained by Thailand.

Importer (进口商) (jìn kǒu shāng)

Any person under the jurisdiction of the State of import who arranges for hazardous wastes or other wastes to be imported. [Basel Convention, Art. 2, No. 16 (1989)]

Improvement of Energy-saving Technology (节能技术进步) (jié néng jì shù jìn bù)

The design and other measures by which technologies are refined or developed to make more efficient use of a specified source of energy.

In-situ Conditions (原地条件) (yuán dì tiáo jiàn)

Conditions where genetic resources exist within ecosystems and natural habitats, and, in the case of domesticated or cultivated species, in the surroundings where they have developed their distinctive properties. [Convention on Biological Diversity, Art. 2 (1992)]

In-situ Conservation (就地保护) (jiù dì bǎo hù)

The conservation of ecosystems and natural habitats and the maintenance and recovery of viable populations of species in their natural surroundings and, in the case of domesticated or cultivated species, in the surroundings where they have developed their distinctive properties. [Convention on Biological Diversity, Art. 2 (1992)] The measures taken to preserve animals and plants in locations and sites within their natural or historic habitat or range (*in situ* is Latin, meaning maintained "in the site").

Incentive Policy for Energy Saving (节能激励措施) (jié néng jī lì cuò shī)

Procedures and practices, established by governmental agencies that regulate or the enterprises that supply energy, to induce users of energy to do so more efficiently by rewarding their prospective undertakings with economic or other benefits.

Incentive Zoning (激励性分区) (jī lì xìng fēn qū)

A system by which zoning incentives are provided to developers on the condition that specific physical, social, or cultural benefits are provided to the community. Incentives include increases in the permissible number of residential units or gross square footage of the developments, or waiver of the height, setback, use, or area provisions of the zoning ordinance. The benefits to be provided in exchange may include affordable housing, recreational facilities, open space, day-care facilities, infrastructure, or cash in lieu thereof. [Nolon, Well Grounded, p. 450 (ELI 2001)]

Incident (偶然事件) (ǒu rán shì jiàn)

Any unintended event, including operating errors, equipment failures, initiating events, accident precursors, near misses or other mishaps, or unauthorized acts, malicious or non-malicious, the consequences or potential consequences of which are not negligible from the point of view of protection or safety. [IAEA Safety Glossary, p. 93 (2007)]

Incineration at Sea (海上焚烧) (hǎi shàng fén shāo)

(1) The disposal of waste by burning at sea on specially designed incinerator ships. Ocean incineration includes the burning of organochlorine compounds and other toxic wastes that are difficult to dispose of. [Glossary of Env't Statistics (1997)] (2) The burning of wastes from vessels off-shore in the marine environment, a practice prohibited for States that have adhered to the 1972 London Convention on the Prevention of Marine Pollution by Dumping of Wastes and Other Matter and a number of waste conventions and regional ocean agreements.

Independent Power Generation System of Renewable Energy (可再生能源独立发电系统) (kě zài shēng néng yuán dú lì fā diàn xì tong)

The processes established for buildings or factories or other entities whereby they produce their own electricity directly, on or near their sites, by use of solar panels, wind turbines, geo-thermal, or hydropower, and do not need to purchase this electricity from other parties.

India's Bhopal Pesticide Spill

See **Bhopal, India Contamination**.

Indian Country (印第安地区) (yìn dì ān dì qū)

Defined at 40 Code of Federal Regulations 122.2 to mean: (1) All land within the limits of any Indian reservation under the jurisdiction of the United States government, notwithstanding the issuance of any patent, and, including rights-of-way running through the reservation; (2) All dependent Indian communities within the borders of the United States whether within the originally or subsequently acquired territory thereof, and whether within or without the limits of a state; and (3) All Indian allotments, the Indian titles to which have not been extinguished, including rights-of-ways running through the same. [US EPA Glossary of NPDES Terms (2004)]

Indian Ocean (印度洋) (yìn dù yáng)

A body of water bound by the Indian Subcontinent, Arabian Peninsula, eastern Africa, Indochina, the Sunda Islands, Australia, and the Southern Ocean.

Indicator Compounds (指标性化合物) (zhǐ biāo xìng huà hé wù)

Chemical compounds, such as carbon dioxide, whose presence at certain concentrations may be used to estimate certain building conditions (e.g., airflow, presence of sources). [US EPA Glossary of Indoor Air Quality Terms (2011)]

Indirect Discharge (间接排放) (jiàn jiē pái fàng)

The introduction of pollutants into a municipal sewage treatment system from any nondomestic source (i.e., any industrial or commercial facility) regulated under Section 307(b), (c), or (d) of the CWA. [US EPA Glossary of NPDES Terms (2004)]

Individually or Jointly (单独或共同) (dān dú huò gòng tóng)

In "joint and several" liability, when one or more actors, such as persons or companies or other entities, are responsible for the same conduct, the liability for harm caused by their conduct exists for each individual and also for all actors taken together, so that when one individual actor may be obligated to make financial compensation for victims of the harm, all other actors have a duty to contribute their fair share to the one actor who makes the compensation payment.

Indoor Air Pollutant (室内空气污染物) (shì nèi kōng qì wū rǎn wù)

Particles and dust, fibers, mists, bioaerosols, and gases or vapors. [US EPA Glossary of Indoor Air Quality Terms (2011)]

Indoor Air Quality (IAQ) (室内空气质量) (shì nèi kōng qì zhì liàng)

Indoor air quality is a term referring to the air quality within and around buildings and structures, especially as it relates to the health and comfort of building occupants. [US EPA Glossary of Indoor Air Quality Terms (2011)]

Industrial Energy Saving (工业节能) (gōng yè jié néng)

The practices and standards for industrial facilities to use energy as efficiently as possible, in a continuously upgraded manner.

Industrial Policies for Energy Saving (节能产业政策) (jié néng chǎn yè zhèng cè)

The norms and best practices required by government agencies or adopted by associations of industries, or established by an individual industrial enterprise, to guide the use of energy as efficiently as possible, and to continuously upgrade and improve energy efficiency.

Industrial Rationalization (工业合理化) (gōng yè hé lǐ huà)

The transfer of all or a portion of the calculated level of production of one Party to another, for the purpose of achieving economic efficiencies or responding to anticipated shortfalls in supply as a result of plant closures. [Montreal Protocol, Art. 1, No.8 (1987)]

Industrial Sources (工业排放源) (gōng yè pái fàng yuán)

Non-municipal or industrial sources often generate wastewater that is discharged to surface waters. The types of wastewaters generated at a facility depend on the specific activities undertaken at a particular site, and may include manufacturing or process wastewaters, cooling waters, sanitary wastewater, and stormwater runoff. [US EPA Glossary of NPDES Terms (2004)]

Infiltration (渗透) (shèn tòu)

The downward entry of water into the immediate surface of soil or other material, as contrasted with percolation, which is movement of water through soil layers or material. [Urban Soil Primer, p. 72 (2005)]

Information and Technical Cooperation (信息和技术合作) (xìn xī hé jì shù hé zuò)

Collaborative sharing and exchanges of knowledge about practices and technologies, often for building capacity, between two more parties.

Infrared Radiation (红外线辐射) (hóng wài xiàn fù shè)

Radiation emitted by the Earth's surface, the atmosphere and the clouds. It is also known as terrestrial or long-wave radiation. Infrared radiation has a distinctive range of wavelengths ("spectrum") longer than the wavelength of the red color in the visible part of the spectrum. The spectrum of infrared radiation is practically distinct from that of solar or short-wave radiation because of the difference in temperature between the Sun and the Earth-atmosphere system. [US EPA Glossary of Climate Change Terms (2011)]

Infrastructure (基础设施) (jī chǔ shè shī)

[I]ncludes utilities and improvements needed to support development in a community. Among these are water and sewage systems, lighting, drainage, parks, public buildings, roads and transportation facilities, and utilities. [Nolon, Well Grounded, p. 451 (ELI 2001)]

Innocent Passage (无害通过) (wú hài tōng guò)

Passage is innocent so long as it is not prejudicial to the peace, good order or security of the coastal State. Such passage shall take place in conformity with this Convention and with other rules of international law. [Law of the Sea Convention, Art. 19, No. 1 (1982)]

Inspection (检查) (jiǎn chá)

Activities such as measuring, examining, testing, gauging one or more characteristics of a product or service and comparing these with specified requirements to determine conformity. [Von Zharen, ISO 14000, p. 200 (1996)]

Instantaneous Maximum Limit (瞬时最大限值) (shùn shí zuì dà xiàn zhí)

The maximum allowable concentration of a pollutant determined from the analysis of any discrete or composite sample collected, independent of the flow rate and the duration of the sampling event. [US EPA Glossary of NPDES Terms (2004)]

Integrated Management System (for Facilities and Activities) ((设施和活动的)综合管理系统) ((shè shī hé huó dòng de) zōng hé guǎn lǐ xì tǒng)

A single coherent management system in which all the component parts of an organization are integrated to enable the organization's objectives to be achieved. These component parts include the organizational structure, resources and organizational processes. Personnel, equipment and organizational culture, as well as the documented policies and processes, form parts of the management system. The organizational processes have to address the totality of the requirements on the organization, as established by or in, for example, interested parties, IAEA safety standards and other international codes and standards. [IAEA Safety Glossary, p. 99 (2007)]

Integrated Resource Management (资源综合管理) (zī yuán zōng hé guǎn lǐ)

A process that promotes the coordinated development and management of water, land and related resources, in order to maximize the resultant economic and social welfare in an equitable manner without compromising the sustainability of vital ecosystems. [Drafting Legislation for Sustainable Soils: A Guide, p. 91 (2004)]

Interested Parties (利益相关方) (lì yì xiàng guān fāng)

Individuals or groups concerned with or affected by the environmental performance of an organization. [Von Zharen, ISO 14000, p. 200 (1996)]

Intergenerational Equity (代际公平) (dài jì gōng píng)

The fairness that the present generation owes to future generations in terms of sharing resources so that each generation may meet its legitimate needs without preventing other generations from meeting their legitimate needs.

Intergovernmental Panel on Climate Change (IPCC) (政府间气候变化专门委员会) (zhèng fǔ jiān qì hòu biàn huà zhuān mén wěi yuán huì)

The IPCC was established jointly by the United Nations Environment Programme and the World Meteorological Organization in 1988. The purpose of the IPCC is to assess information in the scientific and technical literature related to all significant components of the issue of climate change. The IPCC draws upon hundreds of the world's expert scientists as authors and thousands as expert reviewers. Leading experts on climate change and environmental, social, and economic sciences from some 60 nations have helped the IPCC to prepare periodic assessments of the scientific underpinnings for understanding global climate change and its consequences. With its capacity for reporting on climate change, its consequences, and the viability of adaptation and mitigation measures, the IPCC is also looked to as the official advisory body to the world's governments on the state of the science of the climate change issue. For example, the IPCC organized the development of internationally accepted methods for conducting national greenhouse gas emission inventories. [US EPA Glossary of Climate Change Terms (2011)]

Intermunicipal Agreements (城市间协议) (chéng shì jiān xié yì)

Compacts among municipalities to perform functions together that they are authorized to perform independently. In the land use area, localities may agree to adopt compatible comprehensive plans and ordinances, as well as other land use regulations, and to establish joint planning, zoning, historic preservation, and conservation advisory boards or to hire joint inspection and enforcement officers. [Nolon, Well Grounded, p. 451 (ELI 2001)]

International Atomic Energy Agency (IAEA) (国际原子能机构) (guó jì yuán zǐ néng jī gòu)

An agency that works with United Nations member states and other partners to promote safe, secure, and peaceful nuclear technologies. [Int'l Atomic Energy Agency (2012)]

International Biological Resources Protection Act (国际生物资源保护法) (guó jì shēng wù zī yuán bǎo hù fǎ)

A United States act passed in 1973 mandating several agencies to work together to create a strategy for preserving biological diversity in developing countries.

International Centre For Genetic Engineering And Biotechnology Articles (国际遗传工程和生物技术中心) (guó jì yí chuán gōng chéng hé shēng wù jì shù zhōng xīn)

A center, created as a special United Nations Industrial Development Organization program, to promote international cooperation in research and training of genetic engineering and biotechnology to benefit developing countries.

International Convention for the Conservation of Atlantic Tunas (ICCAT) (保护大西洋金枪鱼国际公约) (bǎo hù dà xī yáng jīn qiāng yú guó jì gōng yuē)

[A treaty opened for signature May 14, 1966 establishing a]n inter-governmental fishery organization responsible for the conservation of tunas and tuna-like species in the Atlantic Ocean and its adjacent seas. [Int'l Comm'n for the Conservation of Atlantic Tunas (2012)]

International Convention for the Protection of New Varieties Of Plants (国际植物新品种保护公约) (guó jì zhí wù xīn pǐn zhǒng bǎo hù gōng yuē)

A treaty opened for signature on December 2, 1961, and revised in 1972, 1978, and 1991. The Convention's objective is to protect new varieties of plants through intellectual property rights for plant breeders in order to encourage development of new plant varieties for the benefit of society. [International Convention for the Protection of New Varieties of Plants]

International Convention on Civil Liability for Oil Pollution (国际油污损害民事责任公约) (guó jì yóu wū sǔn hài mín shì zé rèn gōng yuē)

A treaty signed on November 29, 1969. The objectives of the Convention are to ensure that adequate compensation is available to persons who suffer damage caused by pollution resulting from the escape or discharge of oil from ships, and to standardize international rules and procedures for determining questions of liability and adequate compensation in such areas. [CIESIN Civil Liability]

International Convention on Prevention of Oil Pollution At Sea (国际防止海上石油污染公约) (guó jì fáng zhǐ hǎi shàng shí yóu wū rǎn gōng yuē)

A treaty opened for signature on May 12, 1954, and amended on . . . April [11,] 1962, and . . . October [21,] 1969. The objective of this Convention was to take action to prevent pollution of the sea by oil discharged from ships. [CIESIN Oil Pollution at Sea]

International Convention Relating to Intervention on the High Seas in Cases of Oil Pollution Casualties (国际公海油污事故干预公约) (guó jì gōng hǎi yóu wū shì gù gān yù gōng yuē)

A treaty opened for signature on November 29, 1969. Objectives of this Convention are to enable countries to take action on the high seas in cases of a maritime casualty resulting in danger of oil pollution of sea and coastlines, and to establish that such action would not affect the principle of freedom of the high seas. [CIESIN]

International Court of Justice (国际法院) (guó jì fǎ yuàn)

The principal judicial organ of the United Nations. The Court, which is composed of 15 judges, has a dual role: in accordance with international law, settling legal disputes between States submitted to it by them, and giving advisory opinions on legal matters referred to it by duly authorized United Nations organs and specialized agencies. [Int'l Court of Justice (2012)]

International Environmental Governance (国际环境治理) (guó jì huán jìng zhì lǐ)

The international agreements for States to take decisions with respect to ambient environmental conditions for actions that affect the oceans and atmosphere, and areas beyond national jurisdiction, and for actions that one State may take which affect other States.

International Labour Organisation (国际劳工组织) (guó jì láo gōng zǔ zhī)

An international organization responsible for drawing up and overseeing international labor standards. [Int'l Labour Org. (2012)]

International Maritime Organization (IMO) (国际海事组织) (guó jì hǎi shì zǔ zhī)

The United Nations specialized agency with responsibility for the safety and security of shipping and the prevention of marine pollution by ships. [Int'l Maritime Org. (2012)]

International Oil Pollution Preparedness, Response and Cooperation Convention (OPRC) (国际油污防备、反应和合作公约) (guó jì yóu wū fáng bèi 、 fǎn yìng hé hé zuò gōng yuē)

A treaty opened for signature on November 30, 1990, to develop measures to deal with marine oil pollution incidents within the member States and in cooperation with other countries. [OPRC 1990]

International Organization (国际组织) (guó jì zǔ zhī)

An intergovernmental organization constituted by States to which its member States have transferred competence over matters governed by this Convention, including the competence to enter into treaties in respect of those matters. [Law of the Sea Convention, Annex IX, Art. 1 (1982)]

International Organization for Standardization (国际标准化组织) (guó jì biāo zhǔn huà zǔ zhī)

The world's largest developer and publisher of international standards. [Int'l Org. for Standardization (2012)]

International Renewable Energy Agency (IRENA) (国际可再生能源机构) (guó jì kě zài shēng néng yuán jī gòu)

An international organization working to promote the widespread and increased adoption and sustainable use of all forms or renewable energy. [Int'l Renewable Energy Agency (2012)]

International Standards, Guidelines and Recommendations
(国际标准、指南和建议) (guó jì biāo zhǔn、zhǐ nán hé jiàn yì)

International standards, guidelines and recommendations are: (a) for food safety, the standards, guidelines and recommendations established by the Codex Alimentarius Commission relating to food additives, veterinary drug and pesticide residues, contaminants, methods of analysis and sampling, and codes and guidelines of hygienic practice; (b) for animal health and zoonoses, the standards, guidelines and recommendations developed under the auspices of the International Office of Epizootics; (c) for plant health, the international standards, guidelines and recommendations developed under the auspices of the Secretariat of the International Plant Protection Convention in cooperation with regional organizations operating within the framework of the International Plant Protection Convention; and (d) for matters not covered by the above organizations, appropriate standards, guidelines and recommendations promulgated by other relevant international organizations open for membership to all Members, as identified by the Committee. [Agreement on the Application of Sanitary and Phytosanitary Measures, Annex A, No. 3 (2012)]

International Tropical Timber Agreement (ITTA)
(国际热带木材协定) (guó jì rè dài mù cái xié dìng)

The name of successive treaties beginning with a treaty opened for signature on November 18, 1983, establishing the International Tropical Timber Organization (ITTO), an intergovernmental organization promoting the conservation and sustainable management, use and trade of tropical forest resources Later versions of the treaty (with the same name) were adopted in 1994 and 2006. The 2006 ITTA entered into force in 2011. [ITTO]

International Union for the Conservation of Nature and Natural Resources (IUCN)
(国际自然与自然资源保护联盟, IUCN) (guó jì zì rán yǔ zì rán bǎo hù lián méng)

(1) The International Union for Conservation of Nature and Natural Resources (IUCN) works on biodiversity, climate change, energy, human livelihoods and greening the world economy by supporting scientific research, managing field projects all over the world, and bringing

governments, NGOs [non-governmental organizations], the UN and companies together to develop policy, laws and best practice. [Int'l Union for Conservation of Nature (2012)] (2) The international and inter-governmental hybrid organization, established by the adoption of its Statutes in 1948 as the International Union for the Protection of Nature, to provide cooperation and capacity building to conserve nature, and facilitate joint programs by expert scientists, by international and national NGOs, by governmental agencies and departments, and by States. From 1990–2008 the organization was also known as the World Conservation Union.

Intervention (干预) (gān yù)

Any action intended to reduce or avert exposure or the likelihood of exposure to sources that are not part of a controlled practice or that are out of control as a consequence of an accident. [IAEA Safety Glossary, p. 100 (2007) (Ref. [1].)]

Invalidity of Claims of Sovereignty over the High Seas (在公海上空主权索赔无效) (zài gōng hǎi shàng kōng zhǔ quán suǒ péi wú xiào)

No State may validly purport to subject any part of the high seas to its sovereignty. [Law of the Sea Convention, Art. 89 (1982)]

Inventory Analysis (库存分析) (kù cún fēn xī)

Phase of Life Cycle Assessment involving the compilation and quantification of inputs and outputs for a given product or service system throughout its life cycle. [Von Zharen, ISO 14000 (1996)]

Involved Agency (涉及机构) (shè jí jī gòu)

An agency that has jurisdiction by law to fund, approve, or directly undertake an action but does not have the primary responsibility for the action as does the lead agency under the New York State Environmental Quality Review Act. [Nolon, Well Grounded, p. 451 (ELI 2001)]

IPPC Directive ("Integrated Pollution Prevention And Control Directive") (IPPC指令 (综合污染预防控制指令)) (IPPCzhǐ lìng (zōng hé wū rǎn yù fáng kòng zhì zhǐ lìng))

A directive of the European Union concerning integrated pollution prevention and control. [IPPC Directive (2012)]

Iron Oxide (氧化铁) (yǎng huà tiě)

A chemical compound composed of iron and oxygen.

ISO 14000 (国际标准化组织14000) (guó jì biāo zhǔn huà zǔ zhī 14000)

The formulation designed and promulgated by the International Organization for Standandization (ISO) for an international standard practice for environmental management systems for corporations and other entities, including an independent auditing process for compliance.

ISO14000 Environmental Management Standard (国际标准化组织14000 环境管理标准) (guó jì biāo zhǔn huà zǔ zhī 14000 huán jìng guǎn lǐ biāo zhǔn)

See ISO 14000.

J

Japan: Agricultural Products II (United States v. Japan) (WTO) (WT/DS76) (日本关于限制农产品进口措施案 (II) (WT/DS76)) (rì běn guān yú xiàn zhì nóng chǎn pǐn jìn kǒu cuò shī àn (II) (WT/DS76))

A 1999 WTO Appellate Body decision in favor of the United States regarding Japan's quarantine requirement for certain agricultural products suspected to harbor a pest. The Appellate Body upheld a 1998 WTO Panel report, finding that Japan had not adequately conducted scientific risk assessment to support its quarantine.

Japan Itai-itai Disease Lawsuit (日本痛痛病诉讼案) (rì běn tòng tòng bìng sù sòng àn)

A 1969 case in which plaintiffs prevailed for damages caused by Itai-itai disease (cadmium poisoning) due to contaminated industrial wastewater from a chemical plant. [ICETT, Itai-itai Disease (2010)]

Japan Kumamoto Minamata Disease Lawsuit (日本熊本县水俣病诉讼案) (rì běn xióng běn xiàn shuǐ yǔ bìng sù sòng àn)

A 1969 case brought in Kumamoto for damages caused by minamata disease (mercury poisoning).

Japan Minimata Disease Event (日本水俣病事件) (rì běn shuǐ yǔ bìng shì jiàn)

A poisoning disease of the central nervous system caused by the ingestion of shellfish and fish contaminated by methylmercury compounds released from a chemical plant. It was first discovered in 1956. [Japan Ministry of the Environment, Minimata Disease (2002)]

Japan Niigata Minamata Disease Lawsuit
(日本新泻水俣病诉讼案) (rì běn xīn xiè shuǐ yǔ bìng sù sòng àn)

A 1967 case brought in Niigata in which plaintiffs prevailed for damages caused by Minamata Disease (mercury poisoning).

Japan Tokyo Air Pollution Suit (东京大气污染诉讼案)
(dōng jīng dà qì wū rǎn sù sòng àn)

A 1996 air pollution case brought by patients suffering from air pollution diseases. The 2007 settlement provided free medical care to asthma sufferers in the Tokyo metropolitan area and the establishment of environmental air pollution standards in Tokyo.

Japan Yokkaichi Asthma Lawsuit
(日本四日市哮喘病诉讼案) (rì běn sì rì shì xiào chuǎn bìng sù sòng àn)

A 1967 air pollution lawsuit won by Yokkaichi asthma patients.

Japan's "Agricultural Land Soil Pollution Prevention Law"
(日本农业用地土壤防污法) (rì běn nóng yè yòng dì tǔ rǎng fáng wū fǎ)

A Japanese law enacted in 1970 and amended in 1999.

Japan's "Basic Environmental Law" (日本环境基本法)
(rì běn huán jìng jī běn fǎ)

A Japanese law enacted in 1993 replacing Japan's Basic Law for Pollution Control.

Japan's "Basic Law For Pollution Control"
(日本公害对策基本法) (rì běn gōng hài duì cè jī běn fǎ)

A Japanese pollution control law enacted in 1967.

Johannesburg Declaration on Sustainable Development (约翰内斯堡可持续发展声明) (yuē hàn nèi sī bǎo kě chí xù fā zhǎn shēng míng)

A declaration signed on September 4, 2002, at the Earth Summit on Sustainable Development. The Parties' Declaration agreed to focus on "the worldwide conditions that pose severe threats to the sustainable development of our people, which include: chronic hunger; malnutrition; foreign occupation; armed conflict; illicit drug problems; organized crime; corruption; natural disasters; illicit arms trafficking; trafficking in persons; terrorism; intolerance and incitement to racial, ethnic, religious and other hatreds; xenophobia; and endemic, communicable and chronic diseases, in particular HIV/AIDS, malaria and tuberculosis."

Joint Assessment (合作评估) (hé zuò píng gū)

Cooperative assessments resulting in formal mutual recognition of certifications. [Von Zharen, ISO 14000, p. 200 (1996)]

Judicial Review (司法审查) (sī fǎ shěn chá)

(1) Judicial oversight of administrative decisions. (2) The oversight by the courts of the decisions and processes of local land use agencies. . . [Nolon, Well Grounded, p. 451 (ELI 2001)]

Jurisdictional Defect (司法管辖缺陷) (sī fǎ guǎn xiá quē xiàn)

When a legislative action or land use determination is taken without following a mandated procedure, such as referral to a county planning agency or the conduct of environmental review, the action or determination suffers from a jurisdictional defect and is void. Without following mandated procedures, public bodies do not have jurisdiction to act. [Nolon, Well Grounded, p. 451 (ELI 2001)]

Justification (正当性论证) (zhèng dāng xìng lùn zhèng)

(1) The process of determining whether a practice is, overall, beneficial, as required by the International Commission on Radiological Protection's

System of Radiological Protection, i.e., whether the benefits to individuals and to society from introducing or continuing the practice outweigh the harm (including radiation detriment) resulting from the practice. (2) The process of determining whether a proposed intervention is likely, overall, to be beneficial, as required by the International Commission on Radiological Protection's System of Radiological Protection, i.e., whether the benefits to individuals and to society (including the reduction in radiation detriment) from introducing or continuing the intervention outweigh the cost of the intervention and any harm or damage caused by the intervention. [IAEA Safety Glossary, p. 102 (2007)]

K

Kawasaki Steel Co., Ltd. Chiba Steel Plant Pollution Lawsuit (川崎制铁有限公司千叶制铁所污染诉讼案) (chuān qí zhì tiě yǒu xiàn gōng sī qiān yè zhì tiě suǒ wū rǎn sù sòng àn)

A 1988 lawsuit in which the plaintiffs prevailed against Kawasaki Steel Company for health problems due to air pollution from its steel mill.

Kyoto Protocol (京都议定书) (jīng dū yì dìng shū)

A protocol to the UN Framework Climate Change Convention, opened for signature on December 11, 1997. The Protocol requires developed countries to reduce their GHG emissions below levels specified for each of them in the Treaty. These targets must be met within a five-year time frame between 2008 and 2012, and add up to a total cut in GHG emissions of at least 5% against the baseline of 1990. [UNEP Guide to Climate Neutrality, p. 194 (2008)]

L

La Niña

See **El Niño**.

Land (土地) (tǔ dì)

(1) The terrestrial bio-productive system that comprises soil, vegetation, other biota, and the ecological and hydrological processes that operate within the system. [Convention to Combat Desertification Art. 1, § (e) (1994)] (2) A delineable area of the Earth's terrestrial surface, encompassing all attributes of the biosphere immediately above or below this surface, including those near the surface, the climate, the soil landscape, the ecosystems, the surface hydrology (including shallow lakes, rivers, marshes and swamps), the surface sedimentary layers and associated groundwater reserve, the animal populations, the human settlement pattern and physical results of past and present human activity (terracing, water storage or drainage structures, roads and buildings, etc.). [IUCN Drafting Legislation for Sustainable Soils: A Guide, p. 92 (2004)]

Land and Environment Court (Australia) (土地和环境法院 (澳大利亚)) (tǔ dì he huán jìng fǎ yuàn (ào dà lì yà))

The Land and Environment Court of New South Wales is a specialist superior court of record. Its jurisdiction includes merits review, judicial review, civil enforcement, criminal prosecution, criminal appeals and civil claims about planning, environmental, land, mining and other legislation. [Land & Environment Court, About Us (2011)]

Land Degradation (土地退化) (tǔ dì tuì huà)

Reduction or loss, in arid, semi-arid and dry sub-humid areas, of the biological or economic productivity and complexity of rainfed cropland, irrigated cropland, or range, pasture, forest and woodlands resulting from land uses or from a process or combination of processes, including processes arising from human activities and habitation patterns, such as: soil erosion caused by wind and/or water; . . . deterioration of the physical, chemical and biological or economic properties of soil; and . . . long-term loss of natural vegetation. [Convention to Combat Desertification, Art. 1, § (f) (1994)]

Land Trust (土地信托) (tǔ dì xìn tuō)

A not-for-profit organization, private in nature, organized to preserve and protect the natural and man-made environment by, among other techniques, holding conservation easements that restrict the use of real property. [Nolon, Well Grounded, p. 451 (ELI 2001)]

Land Use Law (土地利用法) (tǔ dì lì yòng fǎ)

Encompasses the full range of laws and regulations that influence or affect the development and conservation of the land. This law is intensely inter-governmental and interdisciplinary. In land use law, there are countless intersections among federal, state, regional, and local statutes. It is signifi-cantly influenced by other legal regimes such as environmental, administrative, and municipal law. [Nolon, Well Grounded, p. 451 (ELI 2001)]

Land Use Regulation (Local) (土地利用法规 (地方)) (tǔ dì lì yòng fǎ guī (dì fāng))

Laws enacted by the local legislature for the regulation of any aspect of land use and community resource protection, including zoning, subdivi-sion, special use permit or site plan regulation, or any regulation that pre-scribes the appropriate use of property or the scale, location, or intensity of development. [Nolon, Well Grounded, p. 451 (ELI 2001)]

Land-locked State (内陆国) (nèi lù guó)

A State which has no seacoast. [Law of the Sea Convention, Art. 124 (1982)]

Landfill (填埋场) (tián mái chǎng)

Land waste disposal site in which waste is generally spread in thin layers, compacted, and covered with a fresh layer of soil each day. [US EPA Glossary of Climate Change Terms (2011)]

Landmark Preservation Law (界标保存法) (jiè biāo bǎo cún fǎ)

A law designating individual historic or cultural landmarks and sites for protection. It controls the alteration of landmarks and regulates some aspects of adjacent development to preserve the landmarks' integrity. [Nolon, Well Grounded, p. 451 (ELI 2001)]

Landslide (滑坡) (huá pō)

The rapid downhill movement of a mass of soil and loose rock. A general term for most types of mass-movement landforms and processes involving the downslope movement of soil and rock materials. Although landslides have many causes, most involve earth materials with low shear strength, ground-water saturation of materials, an interruption of the slope by natural causes or human activities, or a combination of these. [USDA–NRCS Understanding Soil Risks and Hazards, p. 67 (2004)]

Large Concentrated Animal Feeding Operation (Large CAFO) (动物集中饲养大型作业) (dòng wù jí zhōng sì yǎng dà xíng zuò yè)

An AFO is defined as a Large CAFO if it stables or confines as many or more than the numbers of animals specified in any of the following categories: 700 mature dairy cows, whether milked or dry; 1000 veal calves; 1000 cattle other than mature dairy cows or veal calves. Cattle includes but is not limited to heifers, steers, bulls and cow/calf pairs; 2500 swine, each weighing 55 pounds or more; 10000 swine, each weighing less than 55 pounds; 500 horses; 10000 sheep or lambs; 55000 turkeys; 30000 laying hens or broilers, if the AFO uses a liquid manure handling system; 125000 chickens (other than laying hens), if the AFO uses other than a liquid manure handling system; 82000 laying hens, if the AFO uses other than a liquid manure handling system; 30000 ducks, if the AFO uses other than a liquid manure handling system; or 5000 ducks, if the AFO uses a liquid manure handling system [40 CFR 122.23 (b) (4)]. [US EPA Glossary of NPDES Terms (2004)]

Large Construction Activity (大规模建筑活动) (dà guī mó jiàn zhù huó dòng)

Includes clearing, grading, and excavating resulting in a land disturbance that will disturb equal to or more than five acres of land or will disturb less than five acres of total land area but is part of a larger common plan of development or sale that will ultimately disturb equal to or more than five acres. Large construction activity does not include routine maintenance that is performed to maintain the original line and grade, hydraulic capacity, or original purpose of the site. [US EPA Glossary of NPDES Terms (2004)]

Leaching (沥滤) (lì lǜ)

(1) The physical processes that carry soluble substances through soils by a percolating liquid, such as water. (2) The removal of soluble material from soil or other material by percolating water. [Urban Soil Primer 72 (2005)] (3) The process by which soluble materials in the soil, such as salts, nutrients, pesticide chemicals or contaminants, are washed into a lower layer of soil or are dissolved and carried away by water. [USGS, Water Science Glossary of Terms (2011)]

Lead Agency (牵头机构) (qiān tóu jī gòu)

The "involved agency" under the New York State Environmental Quality Review Act that is principally responsible for undertaking, funding, or approving an action. The lead agency is responsible for determining whether an environmental impact statement is required in connection with the action and for the preparation and filing of the statement if one is required. [Nolon, Well Grounded, p. 451 (ELI 2001)]

Lead Auditor (首席审计师) (shǒu xí shěn jì shī)

Auditor leading a specific environmental audit, who meets the criteria specified in ISO 14012. [Von Zharen, ISO 14000, p. 200 (1996)]

Legal Person (法人) (fǎ rén)

Any organization, corporation, partnership, firm, association, trust, estate, public or private institution, group, political or administrative entity or other person designated in accordance with national legislation who or which has responsibility and authority for any action having implications for protection and safety. [IAEA Safety Glossary, p. 105 (2007)]

Life-cycle (生命周期) (shēng mìng zhōu qī)

Consecutive and inter-linked stages of a product or service system, from the extraction of natural resources to the final disposal. [Von Zharen, ISO 14000, p. 201 (1996)]

Life-cycle Assessment (生命周期评估) (shēng mìng zhōu qī píng gū)

Systematic set of procedures for compiling and examining the inputs and outputs of materials and energy and the associated environmental impacts directly attributable to the functioning of a product or service system throughout its life-cycle. [Von Zharen, ISO 14000, p. 201 (1996)]

Likelihood (可能性) (kě néng xìng)

The probability that an act will happen or result in consequences that can be predicted.

Limnology (湖沼学) (hú zhǎo xué)

That branch of hydrology pertaining to the study of lakes and bodies of water. [USGS, Science in Your Watershed (2011)]

Little Tuna (小型金枪鱼) (xiǎo xíng jīn qiāng yú)

Euthynnus alletteratus; Euthynnus affinis. A highly migratory species of fish. [Law of the Sea Convention, Annex I (1982)]

Living Resources (生物资源) (shēng wù zī yuán)

(1) Animals and plants that are alive and of value or potential value to society. Contrast with non-living resources such as minerals and fossil fuels. (2) The natural systems and animals and plants situated within the biosphere of the Earth.

Loam (肥土) (féi tǔ)

Soil material that is 7 to 27% clay particles, 28 to 50% silt particles, and less than 52% sand particles. [Urban Soil Primer, p. 72 (2005)]

Local Board (地方委员会) (dì fāng wěi yuán huì)

See **Reviewing Board**.

Local Law (地方法律) (dì fāng fǎ lǜ)

The highest form of local legislation. The power to enact local laws is granted by the state constitution to local governments. [Nolon, Well Grounded, p. 451 (ELI 2001)]

Local Legislature (地方立法机关) (dì fāng lì fǎ jī guān)

Public body that adopts and amends the comprehensive plan, zoning, and land use regulations, and sometimes retains the authority to issue certain permits or perform other administrative functions. The local legislature of a city is typically called the city council; of a village, the village board of trustees; and of a town, the town board. [Nolon, Well Grounded, p. 452 (ELI 2001)]

Local Limits (地方限度) (dì fāng xiàn dù)

Conditional discharge limits imposed by municipalities upon industrial or commercial facilities that discharge to the municipal sewage treatment system. [US EPA Glossary of NPDES Terms (2004)]

Loess (黄土) (huáng tǔ)

A widespread, homogeneous, commonly non-stratified, porous, friable, slightly coherent, usually highly calcareous, fine-grained blanket deposit, consisting predominantly of silt with subordinate grain sizes ranging from clay to fine sand. [USGS, Mineral Resources Online Spatial Data (2011)]

Long-range Transboundary Air Pollution (长程跨界空气污染) (cháng chéng kuà jiè kōng qì wū rǎn)

Air pollution whose physical origin is situated wholly or in part within the area under the national jurisdiction of one State and which has adverse effects in the area under the jurisdiction of another State at such a distance that it is not generally possible to distinguish the contribution of individual emission sources or groups of sources. [UNEP Judicial Handbook, p. 81 (2005)]

Longwave Radiation (长波辐射) (cháng bō fú shè)

The radiation emitted in the spectral wavelength greater than 4 micrometers corresponding to the radiation emitted from the Earth and atmosphere. It is sometimes referred to as "terrestrial radiation" or "infrared radiation," although somewhat imprecisely. [US EPA Glossary of Climate Change Terms (2011)]

Los Angeles Smog (洛杉矶烟雾事件) (luò shān jī yān wù shì jiàn)

One of two common types of smog. Los Angeles smog, also known as photochemical smog, is the result of combustion sources and sunlight. Los Angeles is a large city in California, United States.

Lot (一幅土地) (yī fú tǔ dì)

A portion of a subdivision, plat, tract, or other parcel of land considered as a unit for the purpose of transferring legal title from one person or entity to another. [Nolon, Well Grounded, p. 452 (ELI 2001)]

Lot Area (建筑区域) (jiàn zhù qū yù)

The total square footage of a horizontal area included within the property lines. Zoning laws typically set a minimum required lot area for building in each zoning district. [Nolon, Well Grounded, p. 452 (ELI 2001)]

Low Dispersible Radioactive Material (低扩散性放射性物质) (dī kuò sàn xìng fàng shè xìng wù zhì)

Either solid radioactive material, or solid radioactive material in a sealed capsule, that has limited dispersibility and is not in powder form. [IAEA Safety Glossary, p. 112 (2007)]

Low Strength (低强度) (dī qiáng dù)

[Soils] The soil is not strong enough to support loads. [Urban Soil Primer, p. 73 (2005)]

Low-tide Elevations (低潮高地) (dī cháo gāo dì)

A low-tide elevation is a naturally formed area of land which is surrounded by and above water at low tide but submerged at high tide. Where a low-tide elevation is situated wholly or partly at a distance not exceeding the breadth of the territorial sea from the mainland or an island, the low-water line on that elevation may be used as the baseline for measuring the breadth of the territorial sea. . . . Where a low-tide elevation is wholly situated at a distance exceeding the breadth of the territorial sea from the mainland or an island, it has no territorial sea of its own. [Law of the Sea Convention, Art. 13 (1982)]

M

Major Facility (主要设施) (zhǔ yào shè shī)

Any NPDES facility or activity classified as such by the Regional Administrator, or in the case of approved state programs, the Regional Administrator in conjunction with the State Director. Major municipal dischargers include all facilities with design flows of greater than one million gallons per day and facilities with EPA/State approved industrial pretreatment programs. Major industrial facilities are determined based on specific ratings criteria developed by EPA/State. [US EPA Glossary of NPDES Terms (2004)]

Malevolence (恶意) (è yì)

An instance of wishing evil to others. See also **Malice**. Often used interchangeably with malice. However, malice relates to acts or the intention to commit acts. Since the term malice has an established usage under the law, it should be preferred if this is what is meant.

Malevolent

Characterized by malevolence; wishing evil to others. [IAEA Safety Glossary, p. 117 (2007)]

Malice (恶意) (è yì)

The intention to do evil. In law, wrongful intention, especially as increasing the guilt of certain offenses. See also **Malevolent**. Malice aforethought, malicious intent: in law, the intention to commit a crime.

Malicious

Characterized by malice; intending or intended to do harm. [IAEA Safety Glossary, p. 117 (2007)]

Managed Resource Protected Area: Protected Area Managed Mainly for the Sustainable Use of Natural Ecosystems (资源保护区：主要为了自然生态系统的可持续利用而设立的保护地) (zī yuán bǎo hù qū: zhǔ yào wéi le zì rán shēng tài xì tǒng de kě chí xù lì yòng ér shè lì de bǎo hù dì)

Area containing predominantly unmodified natural systems, managed to ensure long-term protection and maintenance of biological diversity, while providing at the same time a sustainable flow of natural products and services to meet community needs. [Guidelines for Protected Area Management Categories, p. 19 (1994)]

Management (管理) (guǎn lǐ)

The collection, transport and disposal of hazardous wastes or other wastes, including after-care of disposal sites. [Basel Convention, Art. 2, § 2 (1989)]

Management of Hazardous Substances and Activities in International Law (国际法上的危险物质和活动管理) (guó jì fǎ shàng de wēi xiǎn wù zhì hé huó dòng guǎn lǐ)

The obligations and procedures required by international agreements by which States control the production, transport and handling of hazardous substances.

Management System for Development and Utilization of Renewable Energy (可再生能源开发利用管理制度) (kě zài shēng néng yuán kāi fā lì yòng guǎn lǐ zhì dù)

The processes and programs by which energy is produced and applied from kinetic tidal or flowing water systems, geo-thermal systems, wind generator systems or solar systems.

Management System for Energy Measurement
(能源计量管理制度) (néng yuán jì liáng guǎn lǐ zhì dù)

The metrics, procedures, and processes established by a governmental authority or an enterprise to manage the collection, archiving, reporting and assessment of data associated with the use of energy.

Mandatory Energy Saving Standard (强制性节能标准) (qiáng zhì xìng jié néng biāo zhǔn)

A clear norm or requirement established usually by a governmental authority to require that energy use not exceed a certain level.

Marine Ecosystem (海洋生态系统) (hǎi yáng shēng tài xì tǒng)

The interacting systems of interrelated living and non-living resources found in the estuaries and ocean regions of the Earth.

Marine Environment (海洋环境) (hǎi yáng huán jìng)

All the natural systems associated with the cycles of living entities found in or affecting the oceans of the Earth.

Marine Resources (海洋资源) (hǎi yáng zī yuán)

The living and non-living natural resources found in or under the oceans.

Maritime Casualty (海损) (hǎi sǔn)

A collision of vessels, stranding or other incident of navigation, or other occurrence on board a vessel or external to it resulting in material damage or imminent threat of material damage to a vessel or cargo. [Law of the Sea Convention, Art. 221 (1982)]

Marlins (马林鱼) (mǎ lín yú)

Tetrapturus angustirostris; *Tetrapturus belone*; *Tetrapturus pfluegeri*; *Tetrapturus albidus*; *Tetrapturus audax*; *Tetrapturus georgei*; *Makaira mazara*; *Makaira indica*; *Makaira nigricans*. Several species of highly migratory fish. [Law of the Sea Convention, Annex I (1982)]

Mass-based Standard (重量标准) (zhòng liàng biāo zhǔn)

A discharge limit that is measured in a mass unit such as pounds per day. [US EPA Glossary of NPDES Terms (2004)]

Master Plan (总体规划) (zǒng tǐ guī huà)

A term used synonymously by many to refer to the comprehensive plan. The statutory, official name for the community's written plan for the future is the comprehensive plan. [Nolon, Well Grounded, p. 452 (ELI 2001)]

Material Transfer Agreement (材料转让协定) (cái liào zhuǎn ràng xié dìng)

A set of administrative procedures agreed by the provider and user of genetic resources on how the accessed material could be sourced, used as well as issues of compliance to benefit sharing principles. [UNU IAS Pocket Guide, p. 14 (2007)]

MCS (多种化学品敏感度) (duō zhǒng huà xué pǐn mǐn gǎn dù)

See **Multiple Chemical Sensitivity**.

MDL (方法检出限度) (fāng fá jiǎn chū xiàn dù)

See **Method Detection Limit**.

MEAs (多边环境协定) (duō biān huán jìng xié dìng)

See **Multilateral Environmental Agreements**.

Means of Transport (运输工具) (yùn shū gōng jù)

(1) Railway rolling stock, sea, lake and river craft and road vehicles; (2) where local conditions so require, porters and pack animals. [Law of the Sea Convention, Art. 129 (1982)]

Mechanically Ventilated Crawlspace System (槽隙通风系统) (cáo xì tōng fēng xì tǒng)

A system designed to increase ventilation within a crawlspace, achieve higher air pressure in the crawlspace relative to air pressure in the soil beneath the crawlspace, or achieve lower air pressure in the crawlspace relative to air pressure in the living spaces, by use of a fan. [US EPA Glossary of Indoor Air Quality Terms (2011)]

Mediation (调解) (tiáo jiě)

A voluntary process of negotiations, conducted by a trained mediator who works with all involved parties to identify their true interests and to achieve a resolution that responds effectively and fully to those interests. [Nolon, Well Grounded, p. 452 (ELI 2001)]

Mediation or Conciliation (调停或和解) (tiáo tíng huò hé jiě)

(1) The provision of services by an independent party to facilitate a dialogue between two or more parties to permit them to resolve a dispute or draw them into a harmonious settlement of a disagreement. (2) An alternative dispute resolution (ADR) technique. In mediation, [a] neutral [party] acts in a facilitative rather than an evaluative capacity – facilitating, among other things, the parties' own evaluation of the merits of each other's positions. The mediator . . . does not offer his or her personal views on the substantive merits of any issue in dispute. Mediation may be useful: [w]here a party has a strong emotional investment in or reaction to the case which is inhibiting constructive settlement discussion; . . .

[w]here there is a question as to whether a party or negotiating attorney understands the nature of the claims, defenses, technical questions, or other relevant information; . . . [w]here something just doesn't add up (the case seems like it should settle – if you were in your adversary's shoes, you can't imagine why you wouldn't settle – and yet the other side seems to be holding out for something different); . . . [w]here there are difficult person-alities involved; . . . [w]here you suspect that opposing counsel is trying to work towards agreement but is having difficulty communicating with the client; . . . [w]here you suspect that opposing counsel is not accurately con-veying information to the client or is otherwise an obstacle to constructive settlement discussions; . . . [w]here there appears to be unrealized potential for advancing negotiations through the development of non-monetary settlement components (pollution prevention measures, exceeding compli-ance goals, timing of commitments, recognition, etc.); . . . [w]here a party believes that he/she/it is being treated unfairly; . . . [w]here a party end-lessly procrastinates or outright refuses to come to the table; . . . [w]here a party does not appear to have a realistic view of the case; . . . [w]here the number of parties is great and/or the issues so numerous or complex; and . . . [w]here, despite everything – competent counsel, a good faith desire among all parties to reach settlement, no shortage of information, good communication – there remains a seemingly unbridgeable gap. [US EPA, When & How to Use ADR (2012)]

Medium Concentrated Animal Feeding Operation (Medium CAFO) (动物集中饲养中型作业) (dòng wù jí zhōng sì yǎng zhōng xíng zuò yè)

The term Medium CAFO includes any AFO with the type and number of animals that fall within any of the ranges listed below and which has been defined or designated as a CAFO. An AFO is defined as a Medium CAFO if: the type and number of animals that it stables or confines falls within any of the following ranges: 200 to 699 mature dairy cows, whether milked or dry; 300 to 999 veal calves; 300 to 999 cattle other than mature dairy cows or veal calves. Cattle includes but is not limited to heifers, steers, bulls and cow/calf pairs; 750 to 2499 swine each weighing 55 pounds or more; 3000 to 9999 swine each weighing less than 55 pounds; 150 to 499 horses; 3000 to 9999 sheep or lambs; 16 500 to 54 999 turkeys; 9000 to 29 999 laying hens or broilers, if the AFO uses a liquid manure handling system; 37 500 to 124 999 chickens (other than laying hens), if the AFO uses other than a liquid manure handling system; 25 000 to 81 999 laying hens, if the AFO uses other than a liquid manure handling system; 10 000 to 29999 ducks, if the AFO uses other than a liquid manure handling system; or 1500 to 4999 ducks, if the AFO uses a liquid manure handling system; and [e]ither one of the following conditions are met: [p]ollutants are discharged into waters of the United States through a man-made ditch, flushing system, or other similar man-made device; or [p]ollutants are discharged directly into waters of the United States which originate outside of and pass over, across, or through the facility or otherwise come into direct contact with the animals confined in the operation [40 CFR 122.23 (b) (6)]. [US EPA Glossary of NPDES Terms (2004)]

Member of the Public (公众成员) (gōng zhòng chéng yuán)

In a general sense, any individual in the population except, for protection and safety purposes, when subject to occupational or medical exposure. For the purpose of verifying compliance with the annual dose limit for public exposure, the representative individual in the relevant critical group. [IAEA Safety Glossary, p. 119 (2007)]

Methane (CH4) (甲烷) (jiǎ wán)

A hydrocarbon that is a greenhouse gas with a global warming potential [GWP] most recently estimated at 23 times that of carbon dioxide (CO_2). Methane is produced through anaerobic (without oxygen) decomposition of waste in landfills, animal digestion, decomposition of animal wastes, production and distribution of natural gas and petroleum, coal production, and incomplete fossil fuel combustion. The GWP is from the IPCC's Third Assessment Report (TAR). [US EPA Glossary of Climate Change Terms (2011)]

Method Detection Limit (MDL) (方法检出限度) (fāng fǎ jiǎn chū xiàn dù)

Defined as the minimum concentration of a substance that can be measured and reported with 99 percent confidence that the analyte concentration is greater than zero and is determined from analysis of a sample in a given matrix containing the analyte. [US EPA Glossary of NPDES Terms (2004)]

Metric Ton (公吨) (gōng dùn)

Common international measurement for the quantity of greenhouse gas emissions. A metric ton is equal to 2205 lbs or 1.1 short tons. [US EPA Glossary of Climate Change Terms (2011)]

Mexico Gas Explosion (墨西哥液化气爆炸事件) (mò xī gē yè huà qì bào zhà shì jiàn)

A 1984 explosion at a liquid petroleum gas (LPG) industrial facility in a town near Mexico City that killed six hundred people and injured many more. Also known as the San Juanico Disaster. [UNEP, Management of Industrial Accident Prevention & Preparedness, p. 32 (1996)]

MGD (百万加仑) (bǎi wàn jiā lún)

See **Million Gallons Per Day**.

Migration (迁移) (qiān yí)

(1) Human or animal movement from one area to another. (2) The movement of radionuclides in the environment as a result of natural processes. Most commonly, movement of radionuclides in association with groundwater flow. [IAEA Safety Glossary, pp. 119–20 (2007)]

Million Gallons Per Day (MGD) (百万加仑/日) (bǎi wàn jiā lún/rì)

A unit of flow commonly used for wastewater discharges. One mgd is equivalent to 1.547 cubic feet per second. [US EPA Glossary of NPDES Terms (2004)]

Minimization of Waste (废物减量化) (fèi wù jiǎn liàng huà)

The process of reducing the amount and activity of radioactive waste to a level as low as reasonably achievable, at all stages from the design of a facility or activity to decommissioning, by reducing waste generation and

by means such as recycling and reuse, and treatment, with due consideration for secondary as well as primary waste. Should not be confused with
volume reduction. [IAEA Safety Glossary, p. 120 (2007)]

Mining (Coal) (采煤) (cǎi méi)

Coal mining employs basically the same traditional mining techniques
used in hard rock mining – underground and surface ("strip") mining.
One of the more efficient but environmentally destructive methods for
mining coal involves "strip" mining. This technique is analogous to the
open pit mining techniques used in hard rock mining whereby the soil
and rock above the coal seam are removed to expose the seam. The seam
is then blasted and the coal is scooped up by huge front end loaders or
electric shovels and transported to a coal processing plant. These coal
preparation plants use a variety of physical (e.g., screening) and chemical
(e.g., flotation using high gravity liquids) methods to separate the raw coal
from all of the non-combustible waste rock and minerals (e.g., pyrite).
The coarser waste rock is piled up adjacent to the mined out area and the
finer coal tailings coming from the preparation plant are discharged as a
thick slurry into a man-made impoundment. After coal mining operations
have ceased, the mine is reclaimed by dumping the waste rock into the pit,
regrading the area to approximate the original contours of the land and
then replanting the area using native grasses and trees. [US EPA Glossary
of NPDES Terms (2004)]

Mining (Hardrock) (岩石开采) (yán shí kāi cǎi)

Traditional hardrock mining usually involves digging tunnels and adits
(horizontal entrances into hillsides) to reach lodes of mineral-rich ore.
Traditional underground mining and open pit mining require that the
ore-bearing rock be removed and then put through a milling and extraction plant to extract the desired minerals. After the ore body is exhausted,
miners move on leaving behind huge mounds of finely ground tailings and
coarser waste rock, as well as an underground tunnel complex and/or open
pit. [US EPA Glossary of NPDES Terms (2004)]

Mining (Mountaintop) (开采 (山顶)) (kāi cǎi (shān dǐng))

A relatively new variant of strip-mining technology. This mining technique is common in West Virginia and eastern Kentucky where there is enough topographic relief that is highly dissected such that the adjacent valleys can serve as repositories for overburden (soil and waste rock). Bulldozers remove all topsoil and vegetation from the top of the mountain. The bedrock above the coal is then blasted to break it up for removal. Huge draglines (the bucket can hold 15–20 pickup trucks) then remove the overburden from on top of the coal seam and dump the waste rock ("spoil") into adjacent valleys. Once exposed, the coal seam is then blasted and front end loaders scoop up the coal and load it into huge dump trucks (capable of carrying 100 tons) that haul the raw coal to the coal preparation plant. [US EPA Glossary of NPDES Terms (2004)] Recent scientific studies have shown that mountaintop removal mining causes serious impacts on human health and the environment, particularly on ecosystems and biodiversity, that mitigation practices cannot successfully remediate.

Mining (Non-metals) (开采 (非金属矿)) (kāi cǎi (fēi jīn shǔ kuàng))

Non-metallic minerals include salt, gypsum, potash, phosphate, borax and other minerals used in the chemical industry. Sand and gravel extracted for glass making and construction (highways, buildings, and dams) . . . [are] also regarded as a form of non-metal mining activity. These minerals are extracted primarily using open pit mining techniques, but may also employ underground mining techniques when the deposit is located hundreds or thousands of feet below the surface. Non-metal minerals and associated waste rock are generally of lower toxicity and more manageable than hardrock and coal mining wastes, but still have the potential to create environmental pollution from rainwater running off the piles. [US EPA Glossary of NPDES Terms (2004)]

Minutes (会议纪要) (huì yì jì yào)

(1) Official notes of meetings. (2) The minutes typically cover the important discussions, facts found, and actions taken at a meeting. [Nolon, Well Grounded, p. 452 (ELI 2001)]

Mitigating the Effects of Drought (减缓干旱影响) (jiǎn huǎn gān hàn yǐng xiǎng)

Activities related to the prediction of drought and intended to reduce the vulnerability of society and natural systems to drought as it relates to combating desertification. [Convention to Combat Desertification, Art. 1, § (d) (1994)]

Mitigation (减缓) (jiǎn huǎn)

Policies and measures designed to reduce emissions of greenhouse gases so as to mitigate[/]reduce the effects of climate change. [UNEP: Kick the Habit, A UN Guide to Climate Neutrality, p. 195 (2008)]

Mitigation Conditions (减缓条件) (jiǎn huǎn tiáo jiàn)

Conditions imposed by a reviewing body on a proposed development project or other action to mitigate its adverse impact on the environment. [Nolon, Well Grounded, p. 452 (ELI 2001)]

Mixed Use (混合利用) (hùn hé lì yòng)

In some zoning districts, multiple principal uses are permitted to coexist on a single parcel of land. Such uses may be permitted, for example, in neighborhood commercial districts, where apartments may be developed over retail space. [Nolon, Well Grounded, p. 451 (ELI 2001)]

Mixing Zone (混合区) (hùn hé qū)

An area where an effluent discharge undergoes initial dilution and is extended to cover the secondary mixing in the ambient water body. A mixing zone is an allocated impact zone where water quality criteria can be exceeded as long as acutely toxic conditions are prevented. [US EPA Glossary of NPDES Terms (2004)]

Mobile Sources (移动污染源) (yí dòng wū rǎn yuán)

Motor vehicles, engines, and equipment that move, or can be moved, from place to place. Mobile sources include vehicles that operate on roads and highways ("on-road" or "highway" vehicles), as well as non-road vehicles, engines, and equipment. Examples of mobile sources are cars, trucks, buses, earth-moving equipment, lawn and garden power tools, ships, railroad locomotives, and airplanes. [US EPA Glossary of Mobile Source Emissions Terms (2012)]

Model Building Codes (模范建筑规范) (mó fàn jiàn zhù guī fàn)

The building codes published by the four Model Code Organizations and commonly adopted by state or other jurisdictions to control local construction activity. [US EPA Glossary of Indoor Air Quality Terms (2011)]

Model Code Organizations (模范规范组织) (mó fàn guī fàn zǔ zhī)

Includes the following agencies and the model building codes they promulgate: Building Officials and Code Administrators International, Inc. (BOCA National Building Code/1993 and BOCA National Mechanical Code/1993); International Conference of Building Officials (Uniform Building Code/1991 and Uniform Mechanical Code/1991); Southern Building Code Congress, International, Inc. (Standard Building Code/1991 and Standard Mechanical Code/1991); Council of American Building Officials (CABO One- and Two-Family Dwelling Code/1992 and CABO Model Energy Code/1993). [US EPA Glossary of Indoor Air Quality Terms (2011)]

Monitoring of the Risks or Effects of Pollution (污染风险或影响的监测) (wū rǎn fēng xiǎn huò yǐng xiǎng de jiān cè)

States shall, consistent with the rights of other States, endeavour, as far as practicable, directly or through the competent international organizations, to observe, measure, evaluate and analyse, by recognized scientific methods, the risks or effects of pollution of the marine environment . . . In particular, States shall keep under surveillance the effects of any activities which they permit or in which they engage in order to determine whether these activities are likely to pollute the marine environment. [Law of the Sea Convention, Art. 204 (1982)]

Monsoon (季风) (jì fēng)

[D]erived from the Arabic word "mausim" which means season . . . [In the Indian subcontinent, it is] used . . . to describe a system of alternating winds which blow persistently from the northeast during the northern winter and from the opposite direction, the southwest, during the northern summer. Thus, the term monsoon actually refers solely to a seasonal wind shift, and not to precipitation. Even though the term monsoon was originally defined for the Indian subcontinent, monsoon circulations exist in other locations of the world as well, such as in Europe, Africa, and the west coasts of Chile and the United States. [NOAA, The Monsoon (2012)]

Moratorium (暂停) (zàn tíng)

[S]uspends the right of property owners to obtain development approvals while the local legislature takes time to consider, draft, and adopt land use regulations or rules to respond to new or changing circumstances not adequately dealt with by its current laws. A moratorium is sometimes used by a community just prior to adopting a comprehensive plan or zoning law, or a major addendum thereto. [Nolon, Well Grounded, p. 451 (ELI 2001)]

More Stringent Measures (更严格的措施) (gèng yán gé de cuò shī)

The additional measures, practices or procedures required to be established after existing remedial management systems prove to be incapable of averting a degradation in ambient environmental conditions or failing to make progress toward meeting a specified objective.

Mount Pinatubo (皮纳图博火山) (pí nà tú bó huǒ shān)

A volcano in the Philippine Islands that erupted in 1991. The eruption of Mount Pinatubo ejected enough particulate and sulfate aerosol matter into the atmosphere to block some of the incoming solar radiation from reaching Earth's atmosphere. This effectively cooled the planet from 1992 to 1994, masking the warming that had been occurring for most of the 1980s and 1990s. [US EPA Glossary of Climate Change Terms (2011)]

MSGP (多行业一般性许可证) (duō hang yè yī bān xìng xǔ kě zhèng)

See **Multi-sector General Permit**.

MSW (城市固体废物) (chéng shì gù tǐ fèi wù)

See **Municipal Solid Waste**.

MS4 (市级独立雨水管道系统) (shì jí dú lì yǔ shuǐ guǎn dào xì tǒng)

See **Municipal Separate Storm Sewer System**.

Multi-family Housing (多户型住宅) (duō hù xíng zhù zhái)

Most zoning maps contain districts where multi-family housing is permitted by the zoning law. Under the district regulations, buildings with three or more dwelling units are permitted to be constructed, such as garden apartments or multi-story apartment buildings. [Nolon, Well Grounded, p. 452 (ELI 2001)]

Multilateral Environmental Agreements (MEAs) (多边环境协定) (duō biān huán jìng xié dìng)

(1) Legally binding agreements between several States related to the environment. [Multilateral Environmental Agreement Negotiator's Handbook (2007)] (2) The class of global environmental treaties established by States since 1972 to govern international cooperation to protect the global environment with respect to biological diversity, the climate, the stratospheric ozone layer and other natural systems of the Earth's biosphere.

Multiple Chemical Sensitivity (MCS) (多种化学品敏感度) (duō zhǒng huà xué pǐn mǐn gǎn dù)

A condition in which a person reports sensitivity or intolerance (as distinct from being "allergic") to a number of chemicals and other irritants at very low concentrations. There are different views among medical professionals about the existence, causes, diagnosis, and treatment of this condition. [US EPA Glossary of Indoor Air Quality Terms (2011)]

Multi-sector General Permit (MSGP) (多行业一般性许可证) (duō háng yè yī bān xìng xǔ kě zhèng)

The Stormwater Multi-Sector General Permit (M SGP) authorizes the discharge of stormwater from industrial facilities, consistent with the terms of the permit, in areas of the United States where EPA manages the NPDES permit program. [US EPA Glossary of NPDES Terms (2004)]

Municipal Clerk (市镇秘书) (shì zhèn mì shū)

The public official authorized by the local legislature to keep official records of the legislative and administrative bodies of the locality. Final determination of reviewing boards ordinarily must be filed with the municipal clerk. [Nolon, Well Grounded, p. 452 (ELI 2001)]

Municipal Separate Storm Sewer System (MS4) (市政独立雨水管道系统) (shì zhèng dú lì yǔ shuǐ guǎn dào xì tǒng)

A conveyance or system of conveyances (including roads with drainage systems, municipal streets, catch basins, curbs, gutters, ditches, man-made channels, or storm drains): Owned and operated by a state, city, town, borough, county, parish, district, association, or other public body (created by or pursuant to state law) having jurisdiction over disposal of sewage, industrial wastes, stormwater, or other wastes, including special districts under state law such as a sewer district, flood control district or drainage district, or similar entity, or an Indian tribe or an authorized Indian tribal organization, or a designated and approved management agency under section 208 of the Clean Water Act (CWA) that discharges to waters of the United States; . . . [d]esigned or used for collecting or conveying stormwater; . . . [w]hich is not a combined sewer; and . . . [w]hich is not part of a publicly owned treatment works (POTW) [40 CFR 122.26 (b) (8)]. [US EPA Glossary of NPDES Terms (2004)]

Municipal Solid Waste (MSW) (城市固体废物) (chéng shì gù tǐ fèi wù)

Residential solid waste and some non-hazardous commercial, institutional, and industrial wastes. This material is generally sent to municipal landfills for disposal. [US EPA Glossary of Climate Change Terms (2011)]

Municipal Sources (城市污染源) (chéng shì wū rǎn yuán)

(1) A municipal source is defined as a publicly owned treatment works (POTW) facility that receives primarily domestic sewage from residential and commercial customers. Non-municipal sources include industrial and commercial facilities that are unique with respect to the products and processes present at the facility. [US EPA Glossary of NPDES Terms (2004)] (2) POTWs collect domestic sewage from houses, other sanitary wastewater, and wastes from commercial and industrial facilities. POTWs discharge conventional pollutants, and are covered by secondary treatment standards and state water quality standards. POTWs also produce biosolids during the treatment process. [US EPA Glossary of NPDES Terms (2004)]

Mutually Agreed Terms (共同商定条款) (gòng tóng shāng dìng tiáo kuǎn)

[Genetic Resources] A set of terms and conditions agreed between the provider and user of genetic resources for prospecting purposes. [UNU IAS Pocket Guide, p. 44 (2007)]

N

N₂O (一氧化二氮) (yī yǎng huà èr dàn)

See Nitrous Oxide.

National Environmental Policy Act (NEPA, United States) (国家环境政策法 (美国)) (guó jiā huán jìng zhèng cè fǎ (měi guó))

Enacted in 1970, NEPA requires all federal agencies to assess the environmental impacts of their proposed actions prior to making decisions such as issuing permits or spending federal money. The principal objective of NEPA is to ensure that projects are designed, located, and operated in ways that reduce adverse and increase beneficial environmental impacts for existing and succeeding generations. [US EPA, NEPA (2012)]

National Park: Protected Area Managed Mainly for Ecosystem Protection and Recreation (国家公园: 主要为了保护生态系统和娱乐而设立的保护地) (guó jiā gōng yuán: zhǔ yào wéi le bǎo hù shēng tài xì tǒng hé yú lè ér shè lì de bǎo hù dì)

Natural area of land and/or sea, designated to (a) protect the ecological integrity of one or more ecosystems for present and future generations, (b) exclude exploitation or occupation inimical to the purposes of designation of the area and (c) provide a foundation for spiritual, scientific, educational, recreational and visitor opportunities, all of which must be environmentally and culturally compatible. [Guidelines for Protected Area Management Categories, p. 65 (1994)]

National Pollutant Discharge Elimination System (NPDES) (国家污染物排放消除制度) (guó jiā wū rǎn wù pái fàng xiāo chú zhì dù)

A national program under Section 402 of the Clean Water Act for regulation of discharges of pollutants from point sources to waters of the United

States. Discharges are illegal unless authorized by an NPDES permit. [US EPA Glossary of NPDES Terms (2004)]

National Pretreatment Standard or Pretreatment Standard (国家预处理标准或预处理标准) (guó jiā yù chǔ lǐ biāo zhǔn huò yù chǔ lǐ biāo zhǔn)

Any regulation promulgated by the EPA in accordance with Sections 307(b) and (c) of the CWA that applies to a specific category of industrial users and provides limitations on the introduction of pollutants into publicly owned treatment works. This term includes the prohibited discharge standards under 40 CFR 403.5, including local limits [40 CFR 403.3 (j)]. [US EPA Glossary of NPDES Terms (2004)]

Natural Drainage Class (自然排水等级) (zì rán pái shuǐ děng jí)

Refers to the frequency and the duration of wet periods under conditions similar to those under which the soil formed. [Urban Soil Primer, p. 71 (2005)]

Natural Gas (天然气) (tiān rán qì)

Underground deposits of gases consisting of 50 to 90 percent methane (CH_4) and small amounts of heavier gaseous hydrocarbon compounds such as propane (C_3H_8) and butane (C_4H_{10}). [US EPA Glossary of Climate Change Terms (2011)]

Natural Heritage (自然遗产) (zì rán yí chǎn)

Encompasses the biodiversity of both species and ecological communities, including animals, plants, fungi, and terrestrial and freshwater communities. [UNEP Judicial Handbook, p. 109 (2005)]

Natural Soil Aggregate (碎石土) (suì shí tǔ)

Granules, blocks, prisms, otherwise known as peds. [Urban Soil Primer, p. 71 (2005)]

Naturally Occurring Radioactive Material (Norm) (天然存在的放射性物质) (tiān rán cún zài de fàng shè xìng wù zhì)

Radioactive material containing no significant amounts of radionuclides other than naturally occurring radionuclides. The exact definition of 'significant amounts' would be a regulatory decision. Material in which the activity concentrations of the naturally occurring radionuclides have been changed by a process is included in naturally occurring radioactive material. Naturally occurring radioactive material or NORM should be used in the singular unless reference is explicitly being made to various materials. [IAEA Safety Glossary, p. 126 (2007)]

Negative Declaration (否定的宣告) (fǒu dìng de xuān gào)

A written determination by a lead agency [under New York's SEQRA] that the implementations of the action as proposed will not result in any significant adverse environmental impacts. A "neg dec" concludes the environmental review process for an action. [Nolon, Well Grounded, p. 452 (ELI 2001)]

Negative Pressure (负压力) (fù yā lì)

Condition that exists when less air is supplied to a space than is exhausted from the space, so the air pressure within that space is less than that in surrounding areas. Under this condition, if an opening exists, air will flow from surrounding areas into the negatively pressurized space. [US EPA Glossary of Indoor Air Quality Terms (2011)]

New Discharger (新排放设施) (xīn pái fàng shè shī)

Any building, structure, facility, or installation: [f]rom which there is or may be a discharge of pollutants; ... [t]hat did not commence the discharge of pollutants at that particular site prior to August 13, 1979; . . . [w]hich is not a new source; and . . . [w]hich has never received a finally effective NPDES permit for discharges at that site. [US EPA Glossary of NPDES Terms (2004)]

New Source (新源) (xīn yuán)

Any building, structure, facility, or installation from which there is or may be a discharge of pollutants, the construction of which commenced: [a]fter promulgation of standards of performance under Section 306 of the CWA which are applicable to such source; or [a]fter proposal of standards of performance in accordance with Section 306 of the CWA which are applicable to such source, but only if the standards are promulgated in accordance with Section 306 of the CWA within 120 days of their proposal. Except as otherwise provided in an applicable new source performance standard, a source is a new source if it meets the definition in 40 CFR 122.2; and . . . [i]t is constructed at a site at which no other source is located; or . . . [i]t totally replaces the process or production equipment that causes the discharge of pollutants at an existing source; or . . . [i]ts processes are substantially independent of an existing source at the same site. In determining whether these processes are substantially independent, the Director shall consider such factors as the extent to which the new facility is integrated with the existing plant; and the extent to which the new facility is engaged in the same general type of activity as the existing source. [US EPA Glossary of NPDES Terms (2004)]

New Source Performance Standards (NSPS) (新源绩效标准) (xīn yuán jì xiào biāo zhǔn)

Technology-based standards for facilities that qualify as new sources under 40 CFR 122.2 and 40 CFR 122.29. Standards consider that the new source facility has an opportunity to design operations to more effectively control pollutant discharges. [US EPA Glossary of NPDES Terms (2004)]

Nitrogen Oxides (NOX) (氮氧化物) (dàn yǎng huà wù)

(1) A group of highly reactive gases that contain nitrogen and oxygen in varying amounts. Many of the nitrogen oxides are colorless and odorless. The common pollutant nitrogen dioxide (NO_2) can often be seen

combined with particles in the air as a reddish-brown layer over many urban areas. Nitrogen oxides are formed when the oxygen and nitrogen in the air react with each other during combustion. The formation of nitrogen oxides is favored by high temperatures and excess oxygen (more than is needed to burn the fuel). The primary sources of nitrogen oxides are motor vehicles, electric utilities, and other industrial, commercial, and residential sources that burn fuels. [US EPA Glossary of Mobile Source Emissions Terms (2012)] (2) Gases consisting of one molecule of nitrogen and varying numbers of oxygen molecules. Nitrogen oxides are produced in the emissions of vehicle exhausts and from power stations. In the atmosphere, nitrogen oxides can contribute to formation of photochemical ozone (smog), can impair visibility, and have health consequences; they are thus considered pollutants. [US EPA Glossary of Climate Change Terms (2011)]

Nitrous Oxide (N_2O) (一氧化二氮) (yī yǎng huà èr dàn)

A powerful greenhouse gas with a global warming potential of 296 times that of carbon dioxide (CO_2). Major sources of nitrous oxide include soil cultivation practices, especially the use of commercial and organic fertilizers, fossil fuel combustion, nitric acid production, and biomass burning. The GWP is from the IPCC's Third Assessment Report (TAR). [US EPA Glossary of Climate Change Terms (2011)]

NMVOCs (非甲烷有机挥发物) (fēi jiǎ wán yǒu jī huī fā wù)

See **Non-methane Volatile Organic Compounds**.

No-fault Liability (无过错责任) (wú guò cuò zé rèn)

(1) [A form of liability in which] a party may be [responsible] for [environmental] cleanup even though its actions were not considered improper when it disposed of the wastes. Also known as strict liability. [US GAO, Environmental Liabilities (2005)] (2) The legal standard by which a party is deemed to be liable to pay compensation for damages caused by an action, defined by law, which causes harm to another party, without either party having to establish who may have been at fault or who caused the action.

Non-compliance Mechanisms (不遵守机制) (bù zūn shǒu jī zhì)

The provisions of an international agreement whereby the States Parties to the agreement establish a committee to consult with any State Party that is unable for a time to be in compliance with the obligations and expectations of the agreement, and agree on such capacity building measures or other steps as will be helpful in bringing the State into compliance.

Non-conforming Building (不合格的建筑) (bù hé gé de jiàn zhù)

A building constructed prior to the adoption of the zoning law or zoning amendment which is not in accordance with the dimensional provisions, such as building height or setback requirements, of that law or amendment. [Nolon, Well Grounded, p. 452 (ELI 2001)]

Non-conforming Use (不合格用途) (bù hé gé yòng tú)

A land use that is not permitted by a zoning law but that already existed at the time the zoning law or its amendment was enacted. Most nonconforming uses are allowed to continue but may not be expanded or enlarged. [Nolon, Well Grounded, p. 451 (ELI 2001)]

Nonconformity (不合格) (bù hé gé)

The nonfulfillment of a specified requirement. [Von Zharen, ISO 14000, p. 201 (1996)]

Nonconventional Pollutants (非常规污染物) (fēi cháng guī wū rǎn wù)

All pollutants that are not included in the list of conventional or toxic pollutants in 40 CFR Part 401. Includes pollutants such as chemical oxygen demand (COD), total organic carbon (TOC), nitrogen, and phosphorus. [US EPA Glossary of NPDES Terms (2004)]

Non-fixed Contamination (非固定污染) (fēi gù dìng wū rǎn).

Contamination that can be removed from a surface during routine conditions of transport. [IAEA Safety Glossary, p. 41 (2007)]

Non-methane Volatile Organic Compounds (NMVOCs) (非甲烷有机挥发物) (fēi jiǎ wán yǒu jī huī fā wù)

Organic compounds, other than methane, that participate in atmospheric photochemical reactions. [US EPA Glossary of Climate Change Terms (2011)]

Non-renewable Resource (非再生资源) (fēi zài shēng zī yuán)

Natural resource that cannot be replaced, regenerated or brought back to its original state once it has been extracted. [Von Zharen, ISO 14000, p. 201 (1996)]

NORM (天然存在的放射性物质) (tiān rán cún zài de fàng shè xìng wù zhì)

See **Naturally Occurring Radioactive Material**.

NORM Residue (天然存在的放射性物质残留物) (tiān rán cún zài de fàng shè xìng wù zhì cán liú wù)

Material that remains from a process and comprises or is contaminated by naturally occurring radioactive material (NORM). A NORM residue may or may not be waste. [IAEA Safety Glossary, p. 128 (2007)]

Normal Operation (正常运作) (zhèng cháng yùn zuò)

(1) Operation within specified operational limits and conditions. For a nuclear power plant, this includes startup, power operation, shutting down, shutdown, maintenance, testing and re-fuelling. [IAEA Safety Glossary, p. 145 (2007)] (2) The standard practices that are routinely expected and followed in administering any on-going activity.

Notice (通告) (tōng gào)

Notice requirements are contained in state and local statutes. They spell out the number of days in advance of a public hearing that public notice must be given and the precise means that must be used. These means may include publication in the official local newspaper and mailing or posting notice in prescribed ways. Failure to provide public notice is a jurisdictional defect and may nullify the proceedings. [Nolon, Well Grounded, pp. 451–2 (ELI 2001)]

Notification of Imminent or Actual Damage
(即将发生的损害或实际损害的通知) (jí jiāng fā shēng de sǔn hài huò shí jì sǔn hài de tōng zhī)

When a State becomes aware of cases in which the marine environment is in imminent danger of being damaged or has been damaged by pollution, it shall immediately notify other States it deems likely to be affected by such damage, as well as the competent international organizations. [Law of the Sea Convention, Art. 198 (1982)]

Notified Body (公告机构) (gōng gào jī gòu)

A notified body is a testing organization that has been selected to perform assessment activities for (a) particular directive (s). It is approved by the competent authority of its member state and notified to the European Commission and all other member states. [Von Zharen, ISO 14000, p. 201 (1996)]

Notifying State (通知国) (tōng zhī guó)

The State that is responsible for notifying ... potentially affected States and the IAEA of an event or situation of actual, potential or perceived radiological significance for other States. This includes: [t]he State Party that has jurisdiction or control over the facility or activity (including space objects) in accordance with Article 1 of the Convention on Early Notification of a Nuclear Accident ... ; or [t]he State that initially detects, or discovers evidence of, a transnational emergency, for example by: detecting significant increases in atmospheric radiation levels of unknown origin; detecting contamination in transboundary shipments; discovering a dangerous source that may have originated in another State; or diagnosing medical symptoms that may have resulted from exposure outside the State. [IAEA Safety Glossary, p. 129 (2007)]

NOX (氮氧化物) (dàn yǎng huà wù)

See **Nitrogen Oxides**.

NPDES (国家污染物排放消除制度) (guó jiā wū rǎn wù pái fàng xiāo chú zhì dù)

See **National Pollutant Discharge Elimination System**.

NSPS (新源绩效标准) (xīn yuán jì xiào biāo zhǔn)

See **New Source Performance Standards**.

Nuclear (adjective) (核的 (形容词)) (hé de (xíng róng cí))

Strictly relating to a nucleus; relating to or using energy released in nuclear fission or fusion. The adjective 'nuclear' is used in many phrases to modify a noun that it cannot logically modify. It must be borne in mind that the meaning of such phrases may be unclear. These phrases may therefore be open to misunderstanding, misrepresentation or mistranslation, and their usage may need to be explained. Such phrases include: nuclear accident; nuclear community; nuclear emergency; nuclear facility; nuclear fuel; nuclear incident; nuclear installation; nuclear material; nuclear medicine; [a] nuclear power; nuclear sabotage; nuclear safety; nuclear security; nuclear terrorism; nuclear trafficking; nuclear watchdog; and nuclear weapon. For example, strictly speaking, "nuclear material" primarily means the material of the atomic nucleus. [IAEA Safety Glossary, p. 129 (2007)]

Nuclear Fuel (核燃料) (hé rán liào)

Fissionable nuclear material in the form of fabricated elements for loading into the reactor core of a civil nuclear power plant or research reactor.

Fresh fuel: new fuel or unirradiated fuel, including fuel fabricated from fissionable material recovered by reprocessing previously irradiated fuel. [IAEA Safety Glossary, p. 131 (2007)]

Nuclear Fuel Cycle (核燃料循环) (hé rán liào xún huán)

All operations associated with the production of nuclear energy, including: . . . [m]ining and processing of uranium or thorium ores; . . . [e]nrichment of uranium; . . . [m]anufacture of nuclear fuel; . . . [o]peration of nuclear reactors (including research reactors); . . . [r]eprocessing of spent fuel; . . . [a]ll waste management activities (including decommissioning) relating to operations associated with the production of nuclear energy; . . . [a]ny related research and development activities. [IAEA Safety Glossary, p. 131 (2007)]

Nuclear Test Cases (Australia & New Zealand v. France) (ICJ, 1974) (澳大利亚、新西兰诉法国核试验案) (ào dà lì yà 、 xīn xī lán sù fǎ guó hé shì yàn àn)

International Court of Justice (ICJ) decisions concerning French nuclear tests in the atmosphere of the South Pacific. [ICJ, Nuclear Tests (Australia v. France, New Zealand v. France) (1974)]

O

O HORIZON (O 层, 腐叶和生物残骸层) (O céng, fǔ yè hé shēng wù cán hé céng)

An organic layer of fresh and decaying plant residue. [Urban Soil Primer, p. 72 (2005)]

OAK (橡树) (xiàng shù)

A species of tree or shrub that is part of the *Quercus* genus.

OAU

See **Organization of African Unity**.

Objectives (目标) (mù biāo)

Statements of attainable, quantifiable, intermediate-term achievements that help accomplish goals contained in the comprehensive plan [of a local government]. [Nolon, Well Grounded, p. 453 (ELI 2001)]

Occupational Safety and Health Convention and the Work Environment (职业安全和卫生及工作环境公约) (zhí yè ān quán hé wèi shēng jí gōng zuò huán jìng gōng yuē)

A 1981 International Labour Organization Convention, under the terms of which Members agree to formulate, implement and periodically review a coherent national policy on occupational safety, occupational health and the working environment.

Occupational Safety and Health Administration
(职业安全与卫生署) (zhí yè ān quǎn yǔ wèi shēng shǔ)

The United States Congress created the Occupational Safety and Health Administration (OSHA) to assure safe and healthful working conditions for working men and women by setting and enforcing standards and by providing training, outreach, education and assistance. [OSHA (2012)]

OECD

See **Organization for Economic Cooperation and Development**.

Off-set Policy (抵消政策) (dǐ xiāo zhèng cè)

The norms established to allow a party to continue an activity that is agreed to be harmful to the environment, only on the condition that the party compensates for the harmful activity with an activity that neutralizes or compensates equally in kind for the harm, by eliminating comparable harm experienced elsewhere.

Off-site Conservation (迁地保护) (qiān dì bǎo hù)

A literal translation of *ex-situ* conservation. See *Ex-situ* **Conservation**.

Off-site Recycling (异地再循环) (yì dì zài xún huán)

The removal of used or discarded materials to a place or site where they may be reprocessed for reuse or a new use, so that their constituent materials can be once again made productive for socio-economic purposes.

Official Map (正式地图) (zhèng shì dì tú)

The adopted map of a municipality showing streets, highways, parks, drainage, and other physical features. [It] is final and conclusive with respect to the location and width of streets, highways, drainage systems, and parks shown thereon and is established to conserve and protect public health, safety, and welfare. [Nolon, Well Grounded, p. 453 (ELI 2001)]

Operational Planning (业务规划) (yè wù guī huà)

The procedures for designing and implementing the measures whereby either an on-going or new activity or program is undertaken, so that it can be successfully implemented.

Ordinance (条例) (tiáo lì)

An act of a local legislature taken pursuant to authority specifically delegated to local governments by the state legislature. The power of villages to adopt ordinances was eliminated in 1974 [in New York State]. Technically, therefore, villages do not adopt, amend, or enforce ordinances. Zoning provisions in villages are properly called zoning laws. [Nolon, Well Grounded, p. 453 (ELI 2001)]

Organic Compounds (有机化合物) (yǒu jī huà hé wù)

Chemicals that contain carbon. Volatile organic compounds vaporize at room temperature and pressure. They are found in many indoor sources, including many common household products and building materials. [US EPA Glossary of Indoor Air Quality Terms (2011)]

Organic Material of Natural Origin (天然有机物) (tiān rán yǒu jī wù)

Carbonaceous materials found in animals and plants.

Organic Matter (有机物) (yǒu jī wù)

Plant and animal residue in the soil in various stages of decomposition. [Urban Soil Primer, p. 73 (2005)]

Organization (组织) (zǔ zhī)

Company, operation, firm, enterprise, institution, or association, or part thereof, whether incorporated or not, public or private. [Von Zharen, ISO 14000, p. 201 (1996)]

Organization for Economic Cooperation and Development (OECD) (经济合作与发展组织) (jīng jì hé zuò yǔ fā zhǎn zǔ zhī)

An organization that promotes policies that will improve the economic and social well-being of people around the world. [Org. for Econ. Cooperation & Dev. (2012)]

Organization of African Unity (OAU) (非洲统一组织) (fēi zhōu tǒng yī zǔ zhī)

An organization of African States that aims to promote their unity and solidarity; increase cooperation amongst the peoples of Africa and internationally; and defend their sovereignty and independence. [OAU (2012)]

Organizational Structure (组织结构) (zǔ zhī jié gòu)

The responsibilities, authorities and relationships, arranged in a pattern, through which an organization performs its functions. [Von Zharen, ISO 14000, p. 202 (1996)]

Osaka, Japan International Airport Noise Litigation (日本大阪国际机场噪音诉讼案) (rì běn dà bǎn guó jì jī chǎng zào yīn sù sòng àn)

A 1969 case over airport noise pollution.

OSPAR Convention

See **Convention for the Protection of the Marine Environment of the Northeast Atlantic.**

Our Common Future ("Brundtland Report")
(我们共同的未来 (布伦特兰报告)) (wǒ mén gòng tóng de wèi lái (bù lún tè lán bào gào))

The title of the 1987 report of the United Nations World Commission on Environment and Development, also known after the name of its chairman as the Brundtland Report, which outlined the importance of the concept of sustainable development. The concept was agreed uponin Agenda 21 at the United Nations Conference on Environment and Development held in Rio de Janeiro in 1992.

Outdoor Air Supply (室外空气供应) (shì wài kōng qì gōng yìng)

Air brought into a building from the outdoors (often through the ventilation system) that has not been previously circulated through the system. Also known as "Make-up Air." [US EPA Glossary of Indoor Air Quality Terms (2011)]

Outer Space Treaty (外空条约) (wài kōng tiáo yuē)

Formally known as the "Treaty on Principles Governing the Activities of States in the Exploration and Use of Outer Space," including the Moon and Other Celestial Bodies, this treaty, which forms the basis of international space law, was opened for signature on January 27, 1967, and entered into force October 10, 1967.

Overlay Zone (叠加区) (dié jiā qū)

A zone or district created by the local legislature for the purpose of conserving natural resources or promoting certain types of development. Overlay zones are imposed over existing zoning districts and contain provisions that are applicable in addition to those contained in the zoning law. [Nolon, Well Grounded, p. 453 (ELI 2001)]

Oxidation (氧化) (yǎng huà)

The act or process of another substance combining with oxygen.

Oxidize (氧化) (yǎng huà)

To chemically transform a substance by combining it with oxygen. [US EPA Glossary of Climate Change Terms (2011)]

Oxygen (氧气) (yǎng qì)

A necessary element found in the atmosphere that is odorless, colorless, and tasteless.

Ozone (O3) (臭氧) (chòu yǎng)

(1) Ozone, the triatomic form of oxygen (O_3), is a gaseous atmospheric constituent. In the troposphere, it is created both naturally and by photochemical reactions involving gases resulting from human activities (photochemical smog). In high concentrations, tropospheric ozone can be harmful to a wide range of living organisms. Tropospheric ozone acts as a greenhouse gas. In the stratosphere, ozone is created by the interaction between solar ultraviolet radiation and molecular oxygen (O_2). Stratospheric ozone plays a decisive role in the stratospheric radiative balance. Depletion of stratospheric ozone, due to chemical reactions that may be enhanced by climate change, results in an increased ground-level flux of ultraviolet (UV)-B radiation. [US EPA Glossary of Climate Change Terms (2011)] (2) A gaseous molecule that contains three oxygen atoms (O_3). Ozone can exist either high in the atmosphere, where it shields the Earth against harmful ultraviolet rays from the sun, or close to the ground, where it is the main component of smog. Ground-level ozone is a product of reactions involving hydrocarbons and nitrogen oxides in the presence of sunlight. Ozone is a potent irritant that causes lung damage and a variety of respiratory problems. [US EPA Glossary of Mobile Source Emissions Terms (2012)]

Ozone Depleting Substance (ODS) (消耗臭氧层物质) (xiāo hào chòu yǎng céng wù zhì)

A family of man-made compounds that includes, but . . . [is] not limited to, chlorofluorocarbons (CFCs), bromofluorocarbons (halons), methyl chloroform, carbon tetrachloride, methyl bromide, and hydrochlorofluorocarbons (HCFCs). These compounds have been shown to deplete stratospheric ozone, and therefore are typically referred to as ODSs. [US EPA Glossary of Climate Change Terms (2011)]

Ozone Layer (臭氧层) (chòu yǎng céng)

The layer of ozone that begins approximately 15 km above Earth and thins to an almost negligible amount at about 50 km, shields the Earth from harmful ultraviolet radiation from the sun. The highest natural concentration of ozone (approximately 10 parts per million by volume) occurs in the stratosphere at approximately 25 km above Earth. The stratospheric ozone concentration changes throughout the year as stratospheric circulation changes with the seasons. Natural events such as volcanoes and solar flares can produce changes in ozone concentration, but man-made changes are of the greatest concern. [US EPA Glossary of Climate Change Terms (2011)]

Ozone Precursors (臭氧前体) (chòu yǎng qián tǐ)

Chemical compounds, such as carbon monoxide, methane, non-methane hydrocarbons, and nitrogen oxides, which in the presence of solar radiation react with other chemical compounds to form ozone, mainly in the troposphere. [US EPA Glossary of Climate Change Terms (2011)]

P

Pacific Ocean (太平洋) (tài píng yáng)

A body of water bound by the Americas, Asia, Australia, the Arctic, and the Southern Ocean.

Packaging (包装) (bāo zhuāng)

[Nuclear Safety] The assembly of components necessary to enclose the radioactive contents completely. It may, in particular, consist of one or more receptacles, absorbent materials, spacing structures, radiation shielding and service equipment for filling, emptying, venting and pressure relief; devices for cooling, absorbing mechanical shocks, handling and tie-down, and thermal insulation; and service devices integral to the package. The packaging may be a box, drum or similar receptacle, or may also be a freight container, tank or intermediate bulk container. (From Ref. [2].) [IAEA Safety Glossary, p. 139 (2007)]

Parcel (一幅土地) (yī fú tǔ dì)

A piece of property. See LOT. [Nolon, Well Grounded, p. 453 (ELI 2001)]

Parent Material (母质) (mǔ zhì)

The unconsolidated organic and mineral material in which soil forms. [Urban Soil Primer, p. 73 (2005)]

Particulate Filter (微粒过滤器) (wēi lì guò lù qì)

An anti-pollution device designed to trap particles in diesel exhaust before they can escape into the atmosphere. [US EPA Glossary of Mobile Source Emissions Terms (2012)]

Particulate Matter (PM) (颗粒物) (kē lì wù)

(1) Very small pieces of solid or liquid matter such as particles of soot, dust, fumes, mists or aerosols. The physical characteristics of particles, and how they combine with other particles, are part of the feedback mechanisms of the atmosphere. [US EPA Glossary of Climate Change Terms (2011)] (2) Tiny particles or liquid droplets suspended in the air that can contain a variety of chemical components. Larger particles are visible as smoke or dust and settle out relatively rapidly. The tiniest particles can be suspended in the air for long periods of time and are the most harmful to human health because they can penetrate deep into the lungs. Some particles are directly emitted into the air. They come from a variety of sources such as cars, trucks, buses, factories, construction sites, tilled fields, unpaved roads, stone crushing, and wood burning. Other particles are formed in the atmosphere by chemical reactions. [US EPA Glossary of Mobile Source Emissions Terms (2012)]

Particulate Matter 2.5 (PM2.5) (直径小于2.5微米的颗粒物) (zhí jìng xiǎo yú èr diǎn wǔ wēi mǐ de kē lì wù)

Particles that are less than 2.5 microns in diameter. These particles are often referred to as "PM fine." PM fine particles are so small that they are not typically visible to the naked eye. In the atmosphere, however, they are significant contributors to haze. Smaller particles are generally more harmful to human health because they can penetrate more deeply into the lungs than larger particles. Virtually all particulate matter from mobile sources is PM2.5. [US EPA Glossary of Mobile Source Emissions Terms (2012)]

Parts Per Billion (PPB) (十亿分之) (shí yì fēn zhī)

Number of parts of a chemical found in one billion parts of a particular gas, liquid, or solid mixture. [US EPA Glossary of Climate Change Terms (2011)]

Parts Per Million (PPM) (百万分之) (bǎi wàn fēn zhī)

Number of parts of a chemical found in one million parts of a particular gas, liquid, or solid. [US EPA Glossary of Climate Change Terms (2011)]

Party (缔约方) (dì yuē fāng)

A State or regional economic integration organization that has consented to be bound by this Convention and for which the Convention is in force. [Rotterdam Convention, Art. 2 (1998)]

Passage (通道) (tōng dào)

Navigation through the territorial sea for the purpose of: (a) traversing that sea without entering internal waters or calling at a roadstead or port facility outside internal waters; or (b) proceeding to or from internal waters or a call at such roadstead or port facility. [Law of the Sea Convention, Art. 18, No. 1 (1982)]

Patents (专利权) (zhuān lì quán)

The provision to an inventor of a temporary, limited monopoly during which period the inventor may exploit the invention [free] from any direct competition. Patents on their own do not grant anything. They only provide the legal means by which the inventor can prohibit another from using the invention. Usually these are country specific. [UNU IAS Pocket Guide, p. 14 (2007)]

Percolation (过滤) (guò lǜ)

The movement of water through soil. [Urban Soil Primer, p. 73 (2005)]

Perfluorocarbons (PFCs) (全氟化碳) (quán fú huà tàn)

A group of human-made chemicals composed of carbon and fluorine only. These chemicals (predominantly CF_4 and C_2F_6) were introduced as alternatives, along with hydrofluorocarbons, to the ozone depleting substances. In addition, PFCs are emitted as by-products of industrial processes and are also used in manufacturing. PFCs do not harm the stratospheric ozone layer, but they are powerful greenhouse gases: CF_4 has a global warming potential (GWP) of 5700 and C_2F_6 has a GWP of 11 900. The GWP is from the IPCC's Third Assessment Report (TAR). [US EPA Glossary of Climate Change Terms (2011)]

Period or Era (年代) (nián dài)

A division of geologic time.

Permeability (渗透性) (shèn tòu xìng)

The quality of the soil that enables water or air to move downward through the profile. The rate at which a saturated soil transmits water is accepted as a measure of this quality. In soil physics, the rate is referred to as "saturated hydraulic conductivity." [Urban Soil Primer, p. 73 (2005)]

Permissible Exposure Limits (PELs) (可允许的暴露限值) (kě yǔn xǔ de bào lù xiàn zhí)

Standards set by the Occupational, Safety and Health Administration (OSHA). [US EPA Glossary of Indoor Air Quality Terms (2011)]

Permit (许可) (xǔ kě)

The permission granted by a governmental authority to undertake an activity or program, usually by a written license or authorization or control document.

Permitting Authority (许可当局) (xú kě dāng jú)

The United States Environmental Protection Agency (EPA), a Regional Administrator of EPA, or an authorized representative. Under the Clean Water Act, most states are authorized to implement the NPDES permit program. State Authorization Process describes the process for authorizing states to implement the NPDES permit program. [US EPA Glossary of NPDES Terms (2004)]

Persistence (持久性) (chí jiǔ xìng)

The act, condition or state of continuing or enduring, often beyond the cause that first produced it.

Person (人) (rén)

Any natural or legal person. [Basel Convention, Art. 2, No. 14 (1989)]

Pest- or Disease-free Area (无病虫害地区) (wú bìng chóng hài dì qū)

An area, whether all of a country, part of a country, or all or parts of several countries, as identified by the competent authorities, in which a specific pest or disease does not occur. [Agreement on the Application of Sanitary and Phytosanitary Measures, Annex A, No. 6 (2012)]

Pesticides (杀虫剂) (shā chóng jì)

(1) A pesticide is any substance or mixture of substances intended for: preventing, destroying, repelling, or mitigating any pest. Though often misunderstood to refer only to insecticides, the term pesticide also applies to herbicides, fungicides, and various other substances used to control pests. Under United States law, a pesticide is also any substance or mixture of substances intended for use as a plant regulator, defoliant, or desiccant. [US EPA Pesticides (2012)] (2) The chemical compounds and agents that injure, inhibit, destroy, or prevent the growth of insects, bacteria, algae, fungi, or plants and animals deemed to be noxious.

Petroleum Hydrocarbons (石油烃类) (shí yóu tīng lèi)

Naturally occurring organic compounds, and the primary constituents in oil, gasoline, diesel, and a variety of solvents and penetrating oils.

PH (酸/碱) (suān/jiǎn)

A measure of the hydrogen ion concentration of water or wastewater; expressed as the negative log of the hydrogen ion concentration in mg/l. A pH of 7 is neutral. A pH less than 7 is acidic, and a pH greater than 7 is basic. [US EPA Glossary of NPDES Terms (2004)]

pH Value (酸/碱值) (suān/jiǎn zhí)

A numerical designation of acidity and alkalinity in soil. [Urban Soil Primer, p. 73 (2005)]

Photosynthesis (光合作用) (guāng hé zuò yòng)

The process by which plants take CO_2 from the air (or bicarbonate in water) to build carbohydrates, releasing O_2 in the process. There are several pathways of photosynthesis with different responses to atmospheric CO_2 concentrations. [US EPA Glossary of Climate Change Terms (2011)]

Physical Impact (物理影响) (wù lǐ yǐng xiǎng)

The effects observed when an activity or action adversely affects the natural or built environment.

Picocurie (PCI) (微微居里) (wēi wēi jū lǐ)

A unit for measuring radioactivity, often expressed as picocuries per liter (pCi/L) of air. "Pico (p)" is a metric prefix that means one one-millionth of one one-millionth. A picocurie is one one-millionth of one one-millionth of a Curie (Ci). [US EPA Glossary of Indoor Air Quality Terms (2011)]

Pipelines (管道) (guǎn dào)

Linear conduits or pipes for the transmission and transport of oil, gas, water, waste effluents, or other fluid substances, from one place to another.

Piracy (海盗行为) (hǎi dào xíng wéi)

(a) Any illegal acts of violence or detention, or any act of depredation, committed for private ends by the crew or the passengers of a private ship or a private aircraft, and directed: (i) on the high seas, against another ship or aircraft, or against persons or property on board such ship or aircraft; (ii) against a ship, aircraft, persons or property in a place outside the jurisdiction of any State; (b) any act of voluntary participation in the operation of a ship or of an aircraft with knowledge of facts making it a pirate ship or aircraft; (c) any act of inciting or of intentionally facilitating an act described in subparagraph (a) or (b). [Law of the Sea Convention, Art. 101 (1982)]

Planned Unit Development (计划单元开发) (jì huà dān yuán kāi fā)

An overlay zoning district that permits land developments on several parcels to be planned as single units and to contain both residential dwellings and commercial uses. It is usually available to landowners as a mixed use option to single uses permitted as-of-right by the zoning law. [Nolon, Well Grounded, p. 453 (ELI 2001)]

Plant Nutrients (植物营养) (zhí wù yíng yǎng)

Any elements taken in by a plant that are essential to its growth.

Plat (分区地图) (fēn qū dì tú)

A site plan or subdivision map that depicts the arrangements of buildings, roads, and other services for a development. [Nolon, Well Grounded, p. 453 (ELI 2001)]

Pocket Park (小型公园) (xiǎo xíng gōng yuán)

A relatively small area reserved for recreation or gardening and surrounded by streets or buildings. [Urban Soil Primer, p. 73 (2005)]

Point Source (点源) (diǎn yuán)

Any discernible, confined, and discrete conveyance, including but not limited to, any pipe, ditch, channel, tunnel, conduit, well, discrete fissure, container, rolling stock, concentrated animal feeding operation (CAFO), landfill leachate collection system, vessel or other floating craft from which pollutants are or may be discharged. This term does not include return flows from irrigated agriculture or agricultural stormwater runoff. [US EPA Glossary of NPDES Terms (2004)]

Polar Bear Agreement (保护北极熊协定) (bǎo hù běi jí xióng xié dìng)

Formally known as the International Agreement on the Conservation of Polar Bears, this agreement was signed on November 15, 1973, by Canada, Denmark, Norway, the Union of Soviet Socialist Republics, and the United States. The treaty aims to protect polar bears in the Arctic Region.

Police Power (治安权) (zhì ān quán)

The power that is held by the state to legislate for the purpose of preserving the public health, safety, morals, and general welfare of the people of the state. The authority that localities have to adopt comprehensive plans and zoning and land use regulations is derived from the state's police power. [Nolon, Well Grounded, p. 453 (ELI 2001)]

Political and/or Economic Integration Organization (政治 (和) (或) 经济一体化组织) (zhèng zhì (hé) (huò) jīng jì yī tǐ huà zǔ zhī)

An organization constituted by sovereign States to which its member States have transferred competence in respect of matters governed by this Convention and which has been duly authorized, in accordance with its internal procedures, to sign, ratify, accept, approve, formally confirm or accede to it. [Basel Convention, Art. 2, No. 20 (1989)]

Pollen (花粉) (huā fěn)

The fertilizing element of flowering plants, consisting of fine, powdery, yellowish grains or spores, sometimes in masses.

Pollutant (Conservative) (长效污染物) (cháng xiào wū rǎn wù)

Pollutants that do not readily degrade in the environment, and which are mitigated primarily by natural stream dilution after entering receiving bodies of waters. Included are pollutants such as metals. [US EPA Glossary of NPDES Terms (2004)]

Pollutant (Non-conservative) (非长效污染物) (fēi cháng xiào wū rǎn wù)

Pollutants that are mitigated by natural biodegradation or other environmental decay or removal processes in the receiving stream after in-stream mixing and dilution have occurred. [US EPA Glossary of NPDES Terms (2004)]

Pollutant Pathways (污染途径) (wū rǎn tú jìng)

Avenues for distribution of pollutants in a building. HVAC [Heating, Ventilation, Air Conditioning] systems are the primary pathways in most buildings; however all building components interact to affect how air movement distributes pollutants. Also, a term used in the IAQ [Indoor Air Quality] Tools for Schools: IAQ Coordinator's Guide. [US EPA Glossary of Indoor Air Quality Terms (2011)]

Pollutants (Pollution) (污染物) (wū rǎn wù)

Unwanted chemicals or other materials found in the environment. Pollutants can harm human health, the environment, and property. Air pollutants occur as gases, liquid droplets, and solids. Once released into the environment, many pollutants can persist, travel long distances, and move from one environmental medium (e.g., air, water, land) to another. [US EPA Glossary of Mobile Source Emissions Terms (2012)]

Pollute (污染) (wū rǎn)

To make or render impure or unclean. [Massachusetts v. EPA, 549 U.S. at 559]

Polluter (污染者) (wū rǎn zhě)

A party whose actions cause matter or energy to be discharged into the ambient environment so that its nature, location, or volume produces adverse environmental consequences or harm.

Pollution (污染) (wū rǎn)

The presence or matter or energy whose nature, location, or volume produces adverse environmental consequences or harm.

Pollution of the Marine Environment (海洋环境污染) (hǎi yáng huán jìng wū rǎn)

[T]he introduction by man, directly or indirectly, of substances or energy into the marine environment, including estuaries, which results or is likely to result in such deleterious effects as harm to living resources and marine life, hazards to human health, hindrance to marine activities, including fishing and other legitimate uses of the sea, impairment of quality for use of sea water and reduction of amenities. [Law of the Sea Convention, Art. 1, No. 1 (1982)]

Pomfrets (鲳鱼) (chāng yú)

Family *Bramidae*. [Law of the Sea Convention, Annex I (1982)]

Ports (港口) (gǎng kǒu)

For the purpose of delimiting the territorial sea, the outermost permanent harbour works which form an integral part of the harbour system are regarded as forming part of the coast. Off-shore installations and artificial islands shall not be considered as permanent harbour works. [Law of the Sea Convention, Art. 11 (1982)]

Positive Declaration (肯定的宣告) (kěn dìng de xuān gào)

A written determination by a lead agency, under [New York's SEQRA], that the implementation of the action as proposed is likely to have a significant adverse impact on the environment and than an environmental impact statements will be required. Also called a "pos dec." [Nolon, Well Grounded, pp. 453–4 (ELI 2001)]

Positive Pressure (正压) (zhèng yā)

Condition that exists when more air is supplied to a space than is exhausted, so the air pressure within that space is greater than that in surrounding areas. Under this condition, if an opening exists, air will flow from the positively pressurized space into surrounding areas. [US EPA Glossary of Indoor Air Quality Terms (2011)]

PPM (ppm) (百万分之) (bǎi wàn fēn zhī)

Stands for "parts per million" and is the usual measuring unit applied to greenhouse gases because of their relatively small quantities in the atmosphere. 0.0001 per cent is 1 ppm. [UNEP Guide to Climate Neutrality, p. 195 (2008)]

Practical Quantification Limit (PQL) (实际可行的数量限度) (shí jì kě xíng de shù liàng xiàn dù)

The lowest level that can be reliably achieved within specified limits of precision and accuracy during routine laboratory operating conditions. [US EPA Glossary of NPDES Terms (2004)]

Practitioner (从业者) (cóng yè zhě)

Private or governmental sponsored entity which develops, owns, sponsors and/or administers an environmental labeling program. [Von Zharen, ISO 14000, p. 202 (1996)]

Precautionary Approach (风险预防方法) (fēng xiǎn yù fáng fāng fǎ)

The precautionary principle is an approach that provides that where adverse environmental harm may be anticipated, a lack of full knowledge that it will be is not a sufficient reason to delay or defer taking reasonable measures to avert the harm.

Precautionary Principle (风险预防原则) (fēng xiǎn yù fáng yuán zé)

In its most basic form, the [Precautionary Principle] is a strategy to cope with scientific uncertainties in the assessment and management of risks. It is about the wisdom of action under uncertainty: "Look before you leap", "better safe than sorry", and many other folkloristic idioms capture some aspect of this wisdom. Precaution means taking action to protect human health and the environment against possible danger of severe damage. [UNESCO, The Precautionary Principle 8 (2005)]

Precession (岁差) (suì chā)

The comparatively slow torquing of the orbital planes of all satellites with respect to the Earth's axis, due to the bulge of the Earth at the equator which distorts the Earth's gravitational field. Precession is manifest by the slow rotation of the line of nodes of the orbit (westward for inclinations less than 90 degrees and eastward for inclinations greater than 90 degrees). [US EPA Glossary of Climate Change Terms (2011)]

Precipitation (降水) (jiàng shuǐ)

Any form of condensed atmospheric water vapor that falls under gravity.

Preliminary Plat Approval (初步分区地图批准) (chū bù fēn qū dì tú pī zhǔn)

The approval by the authorized local administrative body of a . . . plat. [Nolon, Well Grounded, p. 454 (ELI 2001)]

Prescribed Activity (规定的活动) (guī dìng de huó dòng)

An existing or potentially threatening process for which consent would be required before undertaking the activity. [Drafting Legislation for Sustainable Soils: A Guide, p. 92 (2004)]

Present and Future Generations (当代和后代) (dāng dài hé hòu dài)

The living generation of humans on Earth at any given time and the expectation that this generation will give birth to the next generation. See also **Intergenerational Equity**.

Pressed Wood Products (压木制品) (yā mù zhì pǐn)

A group of materials used in building and furniture construction that are made from wood veneers, particles, or fibers bonded together with an adhesive under heat and pressure. [US EPA Glossary of Indoor Air Quality Terms (2011)]

Pressure, Static (静态压力) (jìng tài yā lì)

In flowing air, the total pressure minus velocity pressure. The portion of the pressure that pushes equally in all directions. [US EPA Glossary of Indoor Air Quality Terms (2011)]

Pressure, Total (总压力) (zǒng yā lì)

In flowing air, the sum of the static pressure and the velocity pressure. [US EPA Glossary of Indoor Air Quality Terms (2011)]

Pressure, Velocity (压力, 速率) (yā lì, sù lǜ)

In flowing air, the pressure due to the velocity and density of the air. [US EPA Glossary of Indoor Air Quality Terms (2011)]

Pretreatment (预处理) (yù chǔ lǐ)

The reduction of the amount of pollutants, the elimination of pollutants, or the alteration of the nature of pollutant properties in wastewater prior to or in lieu of discharging or otherwise introducing such pollutants into a publicly owned treatment works [40 CFR 403.3 (q)]. [US EPA Glossary of NPDES Terms (2004)]

Prevention of Pollution (污染防治) (wū rǎn fáng zhì)

Use of processes, practices, materials, products or energy that avoid or reduce the creation of pollution and waste. [Von Zharen, ISO 14000, p. 202 (1996)]

Preventive Maintenance (预防性维护) (yù fáng xìng wéi hù)

Regular and systematic inspection, cleaning, and replacement of worn parts, materials, and systems. Preventive maintenance helps to prevent parts, material, and systems failure by ensuring that parts, materials and systems are in good working order. [US EPA Glossary of Indoor Air Quality Terms (2011)]

Primary Industry Categories (主要工业分类) (zhǔ yào gōng yè fēn lèi)

Any industry category listed in the Natural Resources Defense Council (NRDC) settlement agreement [NRDC et al. v. Train, 8 E.R.C. 2120 (D.D.C. 1976), modified 12 E.R.C. 1833 (D.D.C. 1979)] for which EPA has or will develop effluent guidelines; also listed in Appendix A of 40 CFR Part 122. [US EPA Glossary of NPDES Terms (2004)]

Primary Treatment (一级处理) (yī jí chǔ lǐ)

The practice of removing some portion of the suspended solids and organic matter in a wastewater through sedimentation. Common usage of this term also includes preliminary treatment to remove wastewater constituents that may cause maintenance or operational problems in the system (i.e., grit removal, screening for rags and debris, oil and grease removal, etc.). [US EPA Glossary of NPDES Terms (2004)]

Principal Use (主要用途) (zhǔ yào yòng tú)

The primary use of a lot that is permitted under the district regulations in a zoning law. These regulations may allow one or more principal uses in any given district. Unless district regulations allow mixed uses, only one principal use may be made of a single lot, along with uses that are accessory to the principal use. [Nolon, Well Grounded, p. 454 (ELI 2001)]

Principle of International Cooperation (国际合作原则) (guó jì hé zuò yuán zé)

The duty of all States to cooperate with each other in good faith for the advancement of sustainable development and the peaceful relations among nations.

Principles of Sustainable Development (可持续发展原则) (kě chí xù fā zhǎn yuán zé)

The Principles contained in the Declaration of Rio de Janeiro on Environmental and Development (1992), and the Johannesburg Declaration on Sustainable Development (2002) and other similar international declarations and agreements; see also **Our Common Future**.

Prior Informed Consent (事先知情同意) (shì xiān zhī qíng tóng yì)

(1) A set of administrative procedures for deciding on whether to grant access to genetic resources on defined terms. [UNU IAS Pocket Guide, p. 15 (2007)] (2) The agreement by a party to an act taken only after another party has given it all relevant information about a proposed action that may have an adverse effect on the first party, and the first party has had time to deliberate and engage its national procedures for public participation, as appropriate, and then determine whether or not to give its agreement to the proposed action.

Priority Pollutants (优先控制的污染物) (yōu xiān kòng zhì de wū rǎn wù)

Those pollutants considered to be of principal importance for control under the CWA based on the NRDC consent decree settlement [NRDC et al. v. Train, 8 E.R.C. 2120 (D.D.C. 1976), modified 12 E.R.C. 1833 (D.D.C. 1979)]; a list of these pollutants is provided as Appendix A to 40 CFR Part 423. [US EPA Glossary of NPDES Terms (2004)]

Procedure (程序) (chéng xù)

A specified way to perform an activity. [Von Zharen, ISO 14000, p. 202 (1996)]

Process (过程) (guò chéng)

A set of interrelated resources and activities which transform inputs into outputs. [Von Zharen, ISO 14000, p. 202 (1996)]

Process Modification (工艺改良) (gōng yì gǎi liáng)

The redesign of a manufacturing process to as to avoid or minimize the production of wastes and use energy and other constituent materials used in manufacturing as efficiently as possible.

Process Wastewater (工艺废水) (gōng yì fèi shuǐ)

Any water which, during manufacturing or processing, comes into direct contact with, or results from the production or use of any raw material, intermediate product, finished product, byproduct, or waste product. [US EPA Glossary of NPDES Terms (2004)]

Product (产品) (chǎn pǐn)

Goods and services for consumer, commercial, and industrial purposes. [Von Zharen, ISO 14000, p. 202 (1996)]

Product Criteria (产品准则) (chǎn pǐn zhǔn zé)

Set of qualitative and quantitative technical requirements that the applicant, product, or product category shall meet to be awarded an environmental label. Product criteria include ecological and product function elements. [Von Zharen, ISO 14000, p. 202 (1996)]

Product Reformulation (产品改造) (chǎn pǐn gǎi zào)

The redesign of a manufactured product so as to avoid the use of harmful components, reduce the production of any wastes associated with the manufacture, and ensure that the product may be recycled to the fullest extent possible.

Production (生产量) (shēng chǎn liàng)

The amount of controlled substances produced, minus the amount destroyed by technologies to be approved by the Parties and minus the amount entirely used as feedstock in the manufacture of other chemicals. The amount recycled and reused is not to be considered as "production." [Montreal Protocol, Art.1, No. 5 (1987)]

Production-based Standard (生产排放标准) (shēng chǎn pái fàng biāo zhǔn)

A discharge standard expressed in terms of pollutant mass allowed in a discharge per unit of product manufactured. [US EPA Glossary of NPDES Terms (2004)]

Prohibition of the Transport of Slaves (奴隶贩运的禁止) (nú lì fàn yùn de jìn zhǐ)

Every State shall take effective measures to prevent and punish the transport of slaves in ships authorized to fly its flag and to prevent the unlawful use of its flag for that purpose. Any slave taking refuge on board any ship, whatever its flag, shall ipso facto be free. [Law of the Sea Convention, Art. 99 (1982)]

Project Cycle (项目周期) (xiàng mù zhōu qī)

The design, production, and implementation phases of a project or undertaking, with appropriate determinations at each phase to ensure sustainable use and environmental protection.

Project Eligibility (项目资格) (xiàng mù zī gé)

An assessment of how a proposed project best meets the design specifications by the party or governmental authority responsible for making that determination.

Propaganda Obligations for Energy-saving (节能宣传义务) (jié néng xuān chuán yì wù)

The duty to undertake public education and communication measures to promote the use of energy-saving practices and technologies.

Proposed Permit (建议的许可证) (jiàn yì de xǔ kě zhèng)

A state NPDES permit prepared after the close of the public comment period (and when applicable, any public hearing and administrative appeals) which is sent to EPA for review before final issuance by the state. [US EPA Glossary of NPDES Terms (2004)]

Protected Area (保护区) (bǎo hù qū)

(1) A geographically defined area which is designated or regulated and managed to achieve specific conservation objectives. [Convention on Biological Diversity, Art. 2 (1992)] (2) An area of land and/or sea especially dedicated to the protection and maintenance of biological diversity, and of natural and associated cultural resources, and managed through legal or other effective means. [Guidelines for Protected Area Management Categories, p. 4 (1994)]

Protected Landscape/Seascape: Protected Area Managed Mainly for Landscape/Seascape Conservation and Recreation (陆地/海洋景观保护区：主要为了陆地/海洋景观保护而设立的保护地) (lù dì/ hǎi yáng jǐng guān bǎo hù qū:zhǔ yào wéi le lù dì/ hǎi yáng jǐng guān bǎo hù ér shè lì de bǎo hù dì)

Area of land, with coast and sea as appropriate, where the interaction of people and nature over time has produced an area of distinct character with significant aesthetic, ecological and/or cultural value, and often with high biological diversity. Safeguarding the integrity of this traditional interaction is vital to the protection, maintenance and evolution of such an area. [Guidelines for Protected Area Management Categories, p. 20 (1994)]

Protection of Human Life (保护人类生命) (bǎo hù rén lèi shēng mìng)

With respect to activities in the Area, necessary measures shall be taken to ensure effective protection of human life. To this end the Authority shall adopt appropriate rules, regulations and procedures to supplement existing international law as embodied in relevant treaties. [Law of the Sea Convention, Art. 146 (1982)]

Protection of Intellectual Property Rights (知识产权保护) (zhī shì chǎn quán bǎo hù)

Legal duties required under laws for copyright, patent, trademark, or other norms for the protection of intellectual property, to ensure that the rights of the property owners are recognized, and royalties are paid as required, and no unauthorized use is undertaken.

Protection of the Marine Environment (保护海洋环境) (bǎo hù hǎi yáng huán jìng)

Necessary measures shall be taken in accordance with this Convention with respect to activities in the Area to ensure effective protection for the marine environment from harmful effects which may arise from such activities. To this end the Authority shall adopt appropriate rules, regulations and procedures for inter alia: (a) the prevention, reduction and control of pollution and other hazards to the marine environment, including the coastline, and of interference with the ecological balance of the marine environment, particular attention being paid to the need for protection from harmful effects of such activities as drilling, dredging, excavation, disposal of waste, construction and operation or maintenance of installations, pipelines and other devices related to such activities; (b) the protection and conservation of the natural resources of the Area and the prevention of damage to the flora and fauna of the marine environment. [Law of the Sea Convention, Art. 145 (1982)]

Protocol Agreement (协议) (xié yì)

An agreement signed between two organizations that operate in different but complementary fields of activity and that commit themselves to take into account their respective assessment results accordingly to conditions specified in advance. [Von Zharen, ISO 14000, p. 202 (1996)]

Protocol on the Implementation of the Alpine Convention of 1991 in the Field of Soil Conservation
(关于实施1991年阿尔卑斯土壤保护公约的议定书)
(guān yú shí shī 1991 niǎn ā ěr bēi sī tú rǎng bǎo hù gōng yūe dě yì dìng shū)

(1) [A 1998 protocol to the 1991 Convention Concerning the Protection of the Alps. Its objectives are:] Safeguarding the multifunctional role of soil based on the concept of sustainable development. To ensure sustainable productivity of soil in its natural function, as an archive of natural and cultural history and in order to guarantee its use for agriculture and forestry, urbanism and tourism, other economic uses, transport and infrastructure, and as a source of raw materials. [Ecolex] (2) Alpine Soil Treaty: The first treaty exclusively dedicated to soil was adopted Oct. 16, 1998 as a protocol to the Nov. 7, 1991 Convention Concerning the Protection of the Alps, 31 I.L.M. 767. The parties recognize that erosion is a problem in the Alpine region because of the topography and as a consequence the concentration of pollutants in the soil can be carried to other ecosystems and present a risk to humans, flora and fauna. [UNEP Judicial Handbook, p. 88 (2005)]

Psychogenic Illness (心理疾病) (xīn lǐ jí bìng)

This syndrome has been defined as a group of symptoms that develop in an individual (or a group of individuals in the same indoor environment) who are under some type of physical or emotional stress. This does not mean that individuals have a psychiatric disorder or that they are imagining symptoms. [US EPA Glossary of Indoor Air Quality Terms (2011)]

Psychosocial Factors (心理因素) (xīn lǐ yīn sù)

Psychological, organizational, and personal stressors that could produce symptoms similar to those caused by poor indoor air quality. [US EPA Glossary of Indoor Air Quality Terms (2011)]

Public Commission (公共委员会) (gōng gòng wěi yuán huì)

A body of individuals convened by a duly authorized authority or governmental agency or intergovernmental body, to convene and deliberate together on a specific topic, usually to make recommendations or take decisions as mandated.

Public Hearing (公开听证会) (gōng kāi tīng zhèng huì)

A hearing that affords citizens affected by a reviewing board's decision an opportunity to have their views heard before decisions are made. [Nolon, Well Grounded, p. 454 (ELI 2001)]

Public Institutions Energy Saving (公共机构节能) (gōng gòng jī gòu jié néng)

The practices and standards for public entities and their buildings to ensure that energy is used efficiently, in accordance with the applicable standards.

Public Institutions Energy Savings Scheme (公共机构节能规划) (gōng gòng jī gòu jié néng guī huà)

Governmental inducements to help public institutions, such as hospitals and schools, invest in energy savings in order to reduce their operating costs and use energy as efficiently as possible.

Public Interest (公众利益) (gōng chòng lì yì)

The best interests of the general community or public, as compared with the private interests of an individual or single enterprise.

Public Prejudice (公妨害) (gōng fáng hài)

Harm or threatened harm to environmental or other amenities broadly enjoyed by the general community or public.

Public Services (公共服务) (gōng gòng fú wù)

Services provided by the municipal government for the benefit of the community, such as fire and police protections, education, solid waste disposal, street cleaning, and snow removal. [Nolon, Well Grounded, p. 454 (ELI 2001)]

Publicly Owned Treatment Works (POTWs) (公有处理厂) (gōng yǒu chǔ lǐ chǎng)

A treatment works, as defined by Section 212 of the CWA, that is owned by the state or municipality. This definition includes any devices and systems used in the storage, treatment, recycling, and reclamation of municipal sewage or industrial wastes of a liquid nature. It also includes sewers, pipes, and other conveyances only if they convey wastewater to a POTW treatment plant [40 CFR 403.3]. Privately-owned treatment works, Federally-owned treatment works, and other treatment plants not owned by municipalities are not considered POTWs. [US EPA Glossary of NPDES Terms (2004)]

**Pulp Mills on the River Uruguay Case (Argentina v. Uruguay)
(乌拉圭河沿岸的纸浆厂案 (阿根廷诉乌拉圭)) (wū lā guī hé yán
àn de zhǐ jiāng chǎng àn (ā gēn tíng sù wū lā guī))**

A 2006 International Court of Justice case regarding the planned and authorized construction of two pulp mills on the River Uruguay.

Purchaser (购买者) (gòu mǎi zhě)

The customer in a contractual situation. [Von Zharen, ISO 14000, p. 203 (1996)]

Q

Qualification Process (资格认定过程) (zī gé rèn dìng guò chéng)

The process of demonstrating whether an entity is capable of fulfilling specified requirements. [Von Zharen, ISO 14000, p. 203 (1996)]

Quartz (石英) (shí yīng)

A mineral consisting of silicon dioxide. Some varieties are semi-precious gemstones.

Quasi-judicial (准司法) (zhǔn sī fǎ)

A term applied to some local administrative bodies that have the power to investigate facts, hold hearings, weigh evidence, draw conclusions, and use this information as a basis for their official decisions. These bodies adjudicate the rights of the parties appearing before the body. [Nolon, Well Grounded, p. 454 (ELI 2001)]

R

R Horizon (R层, 岩石层) (R céng, yán shí céng)

Consolidated soil beneath the soil. The bedrock commonly underlies a C horizon, but it can be directly below an A or a B horizon. [Urban Soil Primer, p. 73 (2005)]

Radiant Heat Transfer (辐射传热) (fú shè chuán rè)

Radiant heat transfer occurs when there is a large difference between the temperatures of two surfaces that are exposed to each other, but are not touching. [US EPA Glossary of Indoor Air Quality Terms (2011)]

Radiation (辐射) (fú shè)

Energy transfer in the form of electromagnetic waves or particles that release energy when absorbed by an object. [US EPA Glossary of Climate Change Terms (2011)]

Radiation Protection Programme (辐射防护计划) (fú shè fáng hù jì huà)

Systematic arrangements which are aimed at providing adequate consideration of radiation protection measures. [IAEA Safety Glossary, p. 156 (2007)]

Radiative Forcing (辐射强迫) (fú shè qiǎng pò)

Radiative forcing is the change in the net vertical irradiance (expressed in Watts per square metre: Wm^{-2}) at the tropopause due to an internal change or a change in the external forcing of the climate system, such as, for example, a change in the concentration of carbon dioxide or the output of the sun. Usually radiative forcing is computed after allowing for stratospheric temperatures to readjust to radiative equilibrium, but with all tropospheric properties held fixed at their unperturbed values. Radiative forcing is called instantaneous if no change in stratospheric temperature is accounted for. Practical problems with this definition, in particular with respect to radiative forcing associated with changes, by aerosols, of the precipitation formation by clouds, are discussed in Chapter 6 of the IPCC Third Assessment Report Working Group I: The Scientific Basis. [US EPA Glossary of Climate Change Terms (2011)]

Radioactive Contents (放射性内容物) (fàng shè xìng nèi róng wù)

The radioactive material together with any contaminated or activated solids, liquids and gases within the packaging. (From Ref. [2].) [IAEA Safety Glossary, p. 158 (2007)]

Radon (Rn) and Radon Decay Products (氡和氡衰变产物) (dōng hé dōng shuāi biàn chǎn wù)

Radon is a radioactive gas formed in the decay of uranium. The radon decay products (also called radon daughters or progeny) can be breathed into the lung where they continue to release radiation as they further decay. [US EPA Glossary of Indoor Air Quality Terms (2011)]

Raised Bed Gardens (园林苗圃床) (yuán lín miáo pǔ chuáng)

Gardens that are planted in boxes made of wood or other materials and have a rooting area above the ground surface. The boxes may be filled with composted materials mixed with uncontaminated soil. [Urban Soil Primer, p. 73 (2005)]

Raw Material (原料) (yuán liào)

Primary or secondary recovered or recycled material that is used in a system to produce a product. [Von Zharen, ISO 14000, p. 203 (1996)]

Re-entrainment (再夹带) (zài jiá dài)

Situation that occurs when the air being exhausted from a building is immediately brought back into the system through the air intake and other openings in the building envelope. [US EPA Glossary of Indoor Air Quality Terms (2011)] See **Re-entry** (synonym).

Re-entry (再入) (zài rù)

Situation that occurs when the air being exhausted from a building is immediately brought back into the system through the air intake and other openings in the building envelope. [US EPA Glossary of Indoor Air Quality Terms (2011)] See **Re-entrainment** (synonym).

Receiving Water (受纳水体) (shòu nà shuǐ tǐ)

The "Water of the United States" as defined in 40 CFR 122.2 into which the regulated stormwater discharges. [US EPA Glossary of NPDES Terms (2004)]

Recommended Energy-saving Standards (推荐性节能标准) (tuī jiàn xìng jié néng biāo zhǔn)

Criteria for the design and use of equipment or activities using energy that are promoted as best practices, and not yet required to be observed by laws.

Recreational Zoning (娱乐分区) (yú lè fēn qū)

The establishment of a zoning district in which private recreational uses are the principal permitted uses. The types of recreational uses permitted include swimming, horseback riding, golf, tennis, exercise clubs open to private members who pay dues and user fees or to the public on a fee basis. [Nolon, Well Grounded, p. 454 (ELI 2001)]

Recycling (回收利用) (huí shōu lì yòng)

(1) Set of processes for reclaiming material that would otherwise be disposed of as a material input to a product or service system. [Von Zharen, ISO 14000, p. 203 (1996)] (2) Collecting and reprocessing a resource so it can be used again. An example is collecting aluminum cans, melting them down, and using the aluminum to make new cans or other aluminum products. [US EPA Glossary of Climate Change Terms (2011)]

Redaction (校订) (jiào dìng)

The process of striking or otherwise taking out of a public record sensitive, private or confidential information, in a way that does not disturb the meaning of the record. [Nolon, Well Grounded, p. 454 (ELI 2001)]

Reefs (礁石) (jiāo shí)

(1) Natural marine structures comprised of the skeletal remains of living organisms, such as coral, and associated sedimentary deposits. (2) In the case of islands situated on atolls or of islands having fringing reefs, the baseline for measuring the breadth of the territorial sea is the seaward low-water line of the reef, as shown by the appropriate symbol on charts officially recognized by the coastal State. [Law of the Sea Convention, Art. 6 (1982)]

Reforestation (重新造林) (chóng xīn zào lín)

Planting of forests on lands that have previously contained forests but that have been converted to some other use. [US EPA Glossary of Climate Change Terms (2011)]

Regime of Islands (岛屿制度) (dǎo yǔ zhì dù)

(1) An island is a naturally formed area of land, surrounded by water, which is above water at high tide. (2) Except as provided for in paragraph 3, the territorial sea, the contiguous zone, the exclusive economic zone and the continental shelf of an island are determined in accordance with the provisions of this Convention applicable to other land territory. (3) Rocks

which cannot sustain human habitation or economic life of their own shall have no exclusive economic zone or continental shelf. [Law of the Sea Convention, Art. 121 (1982)]

Regional Cooperation (区域合作) (qū yù hé zuò)

Agreed upon undertakings by States in the same or continuous geographic regions to collaborate on matters of common interest and concern.

Regional Economic Integration Organization (区域经济一体化组织) (qū yù jīng jì yī tǐ huà zǔ zhī)

An organization constituted by sovereign States of a given region to which its member States have transferred competence in respect of matters governed by this Convention and which has been duly authorized, in accordance with its internal procedures, to sign, ratify, accept, approve or accede to this Convention. [Convention on Biological Diversity, Art. 2 (1992)]; [Convention to Combat Desertification, Art. 1 (1994)]; [Rotterdam Convention, Art. 2 (1998)]

Registration (注册) (zhù cè)

Procedure by which a body indicates relevant characteristics of a product, process or service or particulars of a body or person, and then includes or registers the product. Process or service in an appropriate publicly available list. [Von Zharen, ISO 14000, p. 203 (1996)]

Regulatory Taking (管制性征收) (guǎn zhì xìng zhēng shōu)

A regulation that is so intrusive that it is found to take private property for a public purpose without providing the landowner with just compensation. [Nolon, Well Grounded, p. 454 (ELI 2001)]

Relief (地势) (dì shì)

The elevations or inequalities of a land surface, considered collectively. [Urban Soil Primer, p. 73 (2005)]

RELs (推荐的暴露极限) (tuī jiàn de bào lù jí xiàn)

Recommended Exposure Limits (recommendations made by the National Institute for Occupational Safety and Health (NIOSH)). [US EPA Glossary of Indoor Air Quality Terms (2011)]

Remediation (补救) (bǔ jiù)

Any measures that may be carried out to reduce the radiation exposure from existing contamination of land areas through actions applied to the contamination itself (the source) or to the exposure pathways to humans. Complete removal of the contamination is not implied. The more informal term cleanup is also used. If used, it should be used with the same meaning as remediation, not to attempt to convey a different meaning. The terms rehabilitation and restoration may be taken to imply that the conditions that prevailed before the contamination can be achieved again, which is not normally the case (e.g., owing to the effects of the remedial action itself). Their use is discouraged. [IAEA Safety Glossary, p. 166 (2007)] See **Decontamination**.

Renewable Energy (可再生能源) (kě zài shēng néng yuán)

All forms of energy produced from renewable sources in a sustainable manner. [Statute of the International Renewable Energy Agency, Art. 3 (2009)]

Renewable Energy Law (可再生能源法) (kě zài shēng néng yuán fǎ)

The statute in a State that governs how renewable sources of energy shall be developed and used.

Renewable Energy Law of People's Republic of China (中华人民共和国可再生能源法) (zhōng huá rén mín gòng hé guó kě zài shēng néng yuán fǎ)

A set of laws adopted to promote the development and utilization of renewable energy in the People's Republic of China.

Renewable Energy Technology (可再生能源技术) (kě zaì shēng néng yuán jì shù)

The techniques and equipment appropriate for the design, procurement, and use of different types of renewable energy, as proposed either by national governmental authorities, manufacturers, environmental non-governmental organizations, or intergovernmental international organizations such as the International Renewable Energy Agency (IRENA).

Renewable Resource (可再生资源) (kě zài shēng zī yuán)

Natural resource that is capable of regeneration. [Von Zharen, ISO 14000, p. 203 (1996)]

Requirements of Society (社会要求) (shè huì yāo qiú)

Requirements including laws, statutes, rules and regulations, codes, environmental considerations, health and safety factors, and conservation of energy and materials. [Von Zharen, ISO 14000, p. 203 (1996)]

Reservation of the High Seas for Peaceful Purposes (公海只用于和平目的) (gōng hǎi zh ǐ yòng yú hé píng mù dì)

The high seas shall be reserved for peaceful purposes. [Law of the Sea Convention, Art. 88 (1982)]

Reservoir (库) (kù)

[Climate] [A] component or components of the climate system where a greenhouse gas or a precursor of a greenhouse gas is stored. [United Nations Framework Convention on Climate Change, Art. (1992)]

Residence Time (停留时间) (tíng liú shí jiān)

[Climate] The average time spent in a reservoir by an individual atom or molecule. With respect to greenhouse gases, residence time usually refers to how long a particular molecule remains in the atmosphere. [US EPA Glossary of Climate Change Terms (2011)]

Resolution (决议) (jué yì)

A means by which a local legislature or other board expresses its policy or position on a subject. [Nolon, Well Grounded, p. 454 (ELI 2001)]

Resolution on Incineration at Sea (关于海上焚烧问题的决议) (guān yú hǎi shàng fén shāo wèn tí de jué yì)

A 1994 amendment to the Convention on the Prevention of Marine Pollution by Dumping of Wastes and Other Matter.

Resource Conservation and Recovery Act (RCRA, United States) (资源保护与回收法 (美国)) (zī yuán bǎo hù yǔ huí shōu fǎ (měi guó))

(1) [A federal statute that] sets standards for the treatment, storage, transportation, and disposal of hundreds of different types of hazardous solid waste. A "knowing" violation of these regulations, proven by evidence that the offender was aware of the potential for harm and that he lacked a permit, is punishable by five years imprisonment. Where a person was knowingly placed in danger of serious injury, the offense is punishable by 15 years in prison. [UNEP Judicial Handbook, p. 60 (2005)] (2) Enacted in 1976, it gives EPA the authority to control hazardous waste from the "cradle-to-grave." This includes the generation, transportation, treatment, storage, and disposal of hazardous waste. [US EPA, RCRA (2012)]

Resource Depletion (资源枯竭) (zī yuán kū jié)

Reduction in the global stock of raw material as a result of extraction of non-renewable resources, or extraction of non-renewable resources faster than they can be renewed. [Von Zharen, ISO 14000, p. 203 (1996)]

Resource Survey of Renewable Energy
(可再生能源资源调查) (kě zài shēng néng yuán zī yuán diào chá)

An inventory of the availability of sustained access to particular sources of renewable energy in a specified geographic region.

Resources (资源) (zī yuán)

All solid, liquid or gaseous mineral resources in situ in the Area at or beneath the seabed, including polymetallic nodules; resources, when recovered from the Area, are referred to as "minerals." [Law of the Sea Convention, Art. 133 (1982)]

Respiration (呼吸) (hū xī)

The process whereby living organisms convert organic matter to CO_2, releasing energy and consuming O_2. [US EPA Glossary of Climate Change Terms (2011)]

Response Organization (应急响应组织) (yìng jí xiǎng yìng zǔ zhī)

An organization designated or otherwise recognized by a State as being responsible for managing or implementing any aspect of an emergency response. [IAEA Safety Glossary, p. 168 (2007)]

Response Time (反应时间) (fǎn yìng shí jiān)

The period of time necessary for a component to achieve a specified output state from the time that it receives a signal requiring it to assume that output state. [IAEA Safety Glossary, p. 168 (2007)]

Responsible Care (责任关怀) (zé rèn guān huái)

Comprehensive guidelines for environmental management systems adopted by Chemical Manufacturers Association (CMA) in 1988.

Participation by individual businesses is an obligation of membership in the CMA. [Von Zharen, ISO 14000, p. 204 (1996)]

Restrictive Covenant (限制性约定) (xiàn zhì xìng yuē dìng)

An agreement in writing and signed by the owner of a parcel of land that restricts the use of the parcel in a way that benefits the owners of adjacent or nearby parcels. [Nolon, Well Grounded, p. 454 (ELI 2001)] See **Conservation Easement**.

Restrictive Layer (限制层) (xiàn zhì céng)

A compact, dense layer in a soil that impedes the movement of water and the growth of roots. [Urban Soil Primer, p. 73 (2005)]

Reviewing Board (审查委员会) (shěn chá wěi yuán huì)

The administrative body charged with responsibility for reviewing, approving, conditioning, or denying applications for a specific type of land use such as a variance, special use permit, or site plan or subdivision approval. [Nolon, Well Grounded, p. 454 (ELI 2001)]

Rezoning (再分区) (zài fēn qū)

An act of the local legislature that changes the principal use permitted on one or more parcels of land or throughout one or more zoning districts. Rezoning includes the amendment of the zoning map, and of the use provisions in the district regulations applicable to the land that is rezoned. [Nolon, Well Grounded, p. 455 (ELI 2001)]

Right of Navigation (航行权) (háng xíng quán)

Every State, whether coastal or land-locked, has the right to sail ships flying its flag on the high seas. [Law of the Sea Convention, Art. 90 (1982)]

Rio Declaration on Environment and Development (关于环境与发展的里约热内卢宣言) (guān yú huán jìng yǔ fā zhǎn de lǐ yuē rè nèi lú xuān yán)

The principles of sustainable development adopted by consensus at the 1992 United Nations Conference on Environment and Development in Rio de Janeiro.

Rivers and Harbors Act of 1899 (United States) (1899年河流与港口法) (1899 nían hé liú yǔ gǎng kǒu fǎ)

[A federal statute that] protects the navigable waters of the United States from unauthorized obstructions and refuse. A violation of the act is a misdemeanor punishable by up to one year incarceration. [UNEP Judicial Handbook, p. 60 (2005)]

Roadsteads (外锚地) (wài máo dì)

Roadsteads which are normally used for the loading, unloading and anchoring of ships, and which would otherwise be situated wholly or partly outside the outer limit of the territorial sea, are included in the territorial sea. [Law of the Sea Convention, Art. 12 (1982)]

Root Cause (根源) (gēn yuán)

A fundamental deficiency that results in a nonconformance and must be corrected to prevent recurrence of the same or similar nonconformance. [Von Zharen, ISO 14000 (1996)]

Rotterdam Convention on the Prior Informed Consent Procedure for Certain Hazardous Chemicals and Pesticides in International Trade (关于在国际贸易中对某些危险化学品和农药采用事先知情同意程序的鹿特丹公约) (guān yú zài guó jì mào yì zhōng duì mǒu xiē wéi xiǎn huà xué pǐn hé nóng yào cǎi yòng shì xiān zhī qíng tóng yì chéng xù de lù tè dān gōng yuē)

Also called the Rotterdam Convention, this treaty was opened for signature on September 10, 1998 and entered into force April 24, 2004. This treaty creates legally binding obligations for the implementation of the Prior Informed Consent (PIC) procedure covering pesticides and industrial chemicals that have been banned or severely restricted for health or environmental reasons.

Royalties (专利权使用费) (zhuān lì quán shǐ yòng fèi)

Sources of payment for an inventor who holds the patent for a particular product or process. Royalties accrue when someone uses the patented product or process. [UNU IAS Pocket Guide, p. 15 (2007)]

Rules of Procedure (议事规则) (yì shì guī zé)

The set of norms and specific provisions governing the decision-making and deliberation of a body, such as the UN General Assembly, or the conference of the parties of a multilateral environmental agreement.

Runoff (径流) (jìng liú)

The precipitation discharged into stream channels from an area. The water that flows off the surface of the land without sinking into the soil is called surface runoff. Water that enters the soil before reaching surface streams is called ground-water runoff or seepage flow from ground water. [Urban Soil Primer, p. 73 (2005)]

S

Safe Drinking Water Act (1974) (1974年安全饮用水法) (1974 nián ān quán yǐn yòng shuǐ fǎ)

[A federal statute that] regulates the level of harmful contaminants in public drinking water systems, as well as the underground injection of contaminants into groundwater supplying those systems. A wilful violation of the act is punishable by up to . . . [3] years in prison. [UNEP Judicial Handbook, Ch. 6, Box 15 (2005)]

Salmon and Herring Cases (United States v. Canada) (BISD 35S/98, 1988) (鲑鱼和鲱鱼案 (美国诉加拿大) (BISD 38S/98, 1988)) (guī yú hé fēi yú àn (měi guó sù jiā ná dà))

A 1988 GATT Panel Report regarding Canadian regulations prohibiting the exportation or sale for export of unprocessed herring and pink and sockeye salmon.

Sanitary Sewer (卫生下水道) (wèi shēng xià shuǐ dào)

A pipe or conduit (sewer) intended to carry wastewater or water-borne wastes from homes, businesses, and industries to the POTW. [US EPA Glossary of NPDES Terms (2004)]

Sanitary Sewer Overflows (SSO) (卫生下水道溢出) (wèi shēng xià shuǐ dào yì chū)

Untreated or partially treated sewage overflows from a sanitary sewer collection system. [US EPA Glossary of NPDES Terms (2004)]

Sanitizer (消毒剂) (xiāo dú jì)

One of three groups of antimicrobials registered by EPA for public health uses. EPA considers an anti-microbial to be a sanitizer when it reduces but does not necessarily eliminate all the microorganisms on a treated surface.

To be a registered sanitizer, the test results for a product must show a reduction of at least 99.9% in the number of each test microorganism over the parallel control. [US EPA Glossary of Indoor Air Quality Terms (2011)]

Scoping (确定范围) (què dìng fàn wéi)

A process under [New York's SEQRA] by which the lead agency identifies the potentially significant adverse impacts related to a proposed use and how they are to be addressed in an environmental impact statement (EIS). This process defines the scope of issues to be addressed in the draft EIS, including the content and level of detail of analysis, the range of alternatives, and the mitigation measures needed, as well as issues that do not need to be studied. Scoping provides a project sponsor with guidance on matters that must be considered and provides an opportunity for early participation by involved agencies and the public in the review of the proposal. [Nolon, Well Grounded, p. 455 (ELI 2001)]

Screening (屏蔽) (píng bì)

The act of placing landscape features, such as trees and shrubs, or man-made screens, such as fences or barns, to reduce the impact of development on nearby properties. [Nolon, Well Grounded, p. 455 (ELI 2001)]

Sea Level (海平面) (hǎi píng miàn)

The datum against which land elevation and sea depth are measured. Mean sea level is the average of high and low tides. [NASA Earth Observatory (2012)]

Seabed Mineral Resources (海床矿物资源) (hǎi chuáng kuàng wù zī yuán)

Minerals, whether manganese nodules or hydrocarbons or other, found as resources on and under the seabed.

Seabed (海床) (hái chuáng)

The area beyond the limits of national jurisdiction beneath the oceans of the Earth.

Sealed Soil (密封的土壤) (mì fēng de tǔ rǎng)

Soil that is covered with buildings, pavement, asphalt, or other material. Water and air do not enter the soil from the surface. [Urban Soil Primer, p. 73 (2005)]

Seasonal Migration (季节性迁徙) (jì jié xìng qiān xī)

The movement of animals, including birds, marine mammals, and fish, across their geographic range from season to season.

Secondary Industry Category (辅助产业类) (fǔ zhù chǎn yè lèi)

Any industry category which is not a primary industry category. [US EPA Glossary of NPDES Terms (2004)]

Secondary Treatment (二级处理) (èr jí chǔ lǐ)

Technology-based requirements for direct discharging municipal sewage treatment facilities. Standard is based on a combination of physical and biological processes typical for the treatment of pollutants in municipal sewage. Standards are expressed as a minimum level of effluent quality in terms of: BOD 5, suspended solids (SS), and pH (except as provided for special considerations and treatment equivalent to secondary treatment). [US EPA Glossary of NPDES Terms (2004)]

Security Council (安全理事会) (ān quán lǐ shì huì)

An organization of the United Nations, established by the Charter of the United Nations, with primary responsibility for the maintenance of peace and security.

Sediments (沉积物) (chén jī wù)

Soils, sand, and other minerals washed from land into water bodies and often settling to the bottom of those water bodies.

Self-monitoring (自行监控) (zì xíng jiān kòng)

Sampling and analyses performed by a facility to determine compliance with a permit or other regulatory requirements. [US EPA Glossary of NPDES Terms (2004)]

SEQRA (纽约州环境质量审查法) (niǔ yuē zhōu huán jìng zhì liàng shěn chá fǎ)

New York State Environmental Quality Review Act [SEQRA] requires local legislatures and land use agencies to consider, avoid, and mitigate significant environmental impacts of the project that they approve, the plans or regulations they adopt, and the projects they undertake directly. [Nolon, Well Grounded, p. 455 (ELI 2001)]

Service (服务) (fú wù)

The result generated by activities at the interface between the supplier and the customer and by supplier internal activities to meet the customer needs. [Von Zharen, ISO 14000, p. 204 (1996)]

Service Delivery (服务交付) (fú wù jiāo fù)

Those supplier activities necessary to provide the service. [Von Zharen, ISO 14000, p. 204 (1996)]

Service System (服务系统) (fú wù xì tǒng)

Collection of materially and energetically connected operations which performs one or more defined functions. [Von Zharen, ISO 14000 (1996)]

Setback (缩进) (suō jìn)

A ... restriction [that] requires that no building or structure be located within a specified number of feet from a front, side, or rear lot line. [Nolon, Well Grounded, p. 455 (ELI 2001)]

Severely Hazardous Pesticide Formulation (极为危险的农药制剂) (jí wéi wēi xiǎn de nóng yào zhì jì)

A chemical formulated for pesticidal use that produces severe health or environmental effects observable within a short period of time after single or multiple exposure, under conditions of use. [Rotterdam Convention, Art. 2 (1998)]

Severely Restricted Chemical (严格限用化学品) (yán gé xiàn yòng huà xué pǐn)

A chemical virtually all use of which within one or more categories has been prohibited by final regulatory action in order to protect human health or the environment, but for which certain specific uses remain allowed. It includes a chemical that has, for virtually all use, been refused for approval or been withdrawn by industry either from the domestic market or from further consideration in the domestic approval process, and where there is clear evidence that such action has been taken in order to protect human health or the environment. [Rotterdam Convention, Art. 2 (1998)]

Sewage (污水) (wū shuǐ)

Liquid wastes and wastewater produced by urban settlements or manufacturing or other enterprises, discharged often into the surface waters in areas without sewage treatment facilities.

Sewage Sludge (下水污泥) (xià shuǐ wū ní)

The semi-solid residue remaining from various sewage treatment or other wastewater treatment processes.

Shale (页岩) (yè yán)

Sedimentary rock formed by the hardening of a clay deposit. [Urban Soil Primer, p. 73 (2005)]

Short Ton (美吨) (měi dùn)

Common measurement for a ton in the United States. A short ton is equal to 2000 lbs or 0.907 metric tons. [US EPA Glossary of Climate Change Terms (2011)]

Short-circuiting (短暂空气流通) (duǎn zàn kōng qì liú tōng)

Situation that occurs when the supply air flows to return or exhaust grilles before entering the breathing zone (area of a room where people are). To avoid short-circuiting, the supply air must be delivered at a temperature and velocity that results in mixing throughout the space. [US EPA Glossary of Indoor Air Quality Terms (2011)]

Shrimp–Turtle Case (India, Malaysia, Pakistan, Thailand v. United States) (WT/DS58, 1998) (虾-海龟案) (印度、马来西亚、巴基斯坦、泰国诉美国) (WT/DS58, 1998) (xiā—hǎi guī àn) (yìn dù 、 mǎ lái xī yà 、 bā jī sī tǎn 、 tài guó sù měi guó) (WT/DS58, 1998)

The decision of a 1998 GATT panel regarding a ban by the United States on importation of shrimp and shrimp products from complainant countries. The United States claimed that the complainants did not protect endangered or threatened sea turtles in their harvesting practices. The GATT panel acknowledged that measures to protect sea turtles would be legitimate under GATT Article 20 (i.e., XX), which deals with various exceptions to the WTO's trade rules, but that the United States failed to meet certain criteria such as non-discrimination in enacting the ban.

Shrink–Swell Potential (胀缩潜能) (zhàng suō qiǎn néng)

The potential for volume change in a soil with a loss or gain in moisture. Volume change occurs mainly because of the interaction of clay minerals with water and varies with the amount and type of clay minerals in the soil. The size of the load on the soil and the magnitude of the change in soil moisture content influence the amount of swelling of soils in place. Shrinking and swelling can damage roads, dams, building foundations, and other structures. It can also damage plant roots. [Urban Soil Primer, p. 73 (2005)]

Sick Building Syndrome (SBS) (大楼病综合症) (dà lóu bìng zòng hé zhèng)

Term that refers to a set of symptoms that affect some number of building occupants during the time they spend in the building and diminish or go away during periods when they leave the building. Cannot be traced to specific pollutants or sources within the building. [US EPA Glossary of Indoor Air Quality Terms (2011)]

Sierra Club v. Morton (塞拉俱乐部诉莫顿案) (sāi lā jù lè bù sù mò dùn àn)

A 1972 US Supreme Court case that held that injury to aesthetic, conservation, and recreational interests, as well as to economic interests, could support standing to sue.

Sign Regulation (广告牌规章) (guǎng gào pái guī zhāng)

Local laws that regulate the erection and maintenance of signs and outdoor advertising with respect to their size, color, appearance, movement, and illumination, and their placement on structures or their location on the ground. [Nolon, Well Grounded, p. 455 (ELI 2001)]

Significant Industrial User (SIU) (重要工业用户) (zhòng yào gōng yè yòng hù)

An indirect discharger that is the focus of control efforts under the national pretreatment program; includes all indirect dischargers subject to national categorical pretreatment standards, and all other indirect dischargers that contribute 25 000 gpd [gallons per day] or more of process wastewater, or which make up five percent or more of the hydraulic or organic loading to the municipal treatment plant, subject to certain exceptions [40 CFR 403.3 (t)]. [US EPA Glossary of NPDES Terms (2004)]

Significant Transboundary Release (重大跨界排放) (zhòng dà kuà jiè pái fàng)

A release of radioactive material to the environment that may result in doses or levels of contamination beyond national borders from the release which exceed international intervention levels or action levels for protective actions, including food restrictions and restrictions on commerce. [IAEA Safety Glossary, p. 182 (2007)]

Sink (汇) (huì)

Any process, activity or mechanism which removes a greenhouse gas, an aerosol or a precursor of a greenhouse gas from the atmosphere. [United Nations Framework Convention on Climate Change, Art. 1, No. 8 (1992)]

Sinkhole (岩坑) (yán kēng)

A depression in the landscape where limestone has been dissolved or lava tubes have collapsed. [Urban Soil Primer, p. 73 (2005)]

Site Plan (总平面图) (zǒng píng miàn tú)

[C]onsisting of a map and all necessary supporting material . . . shows the proposed development and use of a single parcel of land. [Nolon, Well Grounded, p. 455 (ELI 2001)]

Slope (坡度) (pō dù)

The inclination of the land surface from the horizontal. Percentage of slope is the vertical distance divided by the horizontal distance, then multiplied by 100. Thus, a slope of 20 percent is a drop of 20 feet in 100 feet of horizontal distance. [Urban Soil Primer, p. 74 (2005)]

Smog (烟雾) (yān wù)

A commonly used term for pollution caused by complex chemical reactions involving nitrogen oxides and hydrocarbons in the presence of sunlight. Ozone is a key component of smog. Smog-forming chemicals come from a wide variety of combustion sources and are also found in products such as paints and solvents. Smog can harm human health, damage the environment, and cause poor visibility. Major smog occurrences are often linked to heavy motor vehicle traffic. [US EPA Glossary of Mobile Source Emissions Terms (2012)]

Soil (土壤) (tǔ rǎng)

(1) The part of the Earth between its surface and its bedrock. It contains the nutrients necessary for maintenance of plant life and it acts to filter out pollutants before they reach subterranean water sources or enter the food chain. Soil also helps to avoid flooding by absorbing considerable amounts of water. Nearly all soil constitutes a habitat for flora and fauna and in this way contributes to biological diversity. In addition to its natural roles, soil is a primary resource for construction, physical support for structures and of historical evidence on the origins of plants, humans, animals and the Earth. [UNEP Judicial Handbook, Ch. 9.1 (2005)]. (2) The natural dynamic system of unconsolidated mineral and organic material at the Earth's surface. It has been developed by physical, chemical and biological processes including the weathering of rock and the decay of vegetation. Soil is an integral part of the Earth's ecosystems and is situated at the interface between the Earth's surface and bedrock. It is subdivided into successive horizontal layers with specific physical, chemical and biological characteristics. From the standpoint of history of soil use, and from an ecological and environmental point of view, the concept of soil also embraces porous sedimentary rocks and other permeable materials together with the water that these contain and the reserves of underground water. [Drafting Legislation for Sustainable Soils: A Guide, p. 92 (2004)]

Soil Carbon (土壤有机碳) (tǔ rǎng yǒu jī tàn)

A major component of the terrestrial biosphere pool in the carbon cycle. The amount of carbon in the soil is a function of the historical vegetative cover and productivity, which in turn is dependent in part upon climatic variables. [US EPA Glossary of Climate Change Terms (2011)]

Soil Conservation (土壤保育) (tǔ rǎng bǎo yù)

Any technical or non-technical process applied to the soil, to ensure that soil, together with all other aspects of the ecosystem, is conserved. [Drafting Legislation for Sustainable Soils: A Guide, p. 92 (2004)]

Soil Conservation Agreement (土壤保育协议) (tǔ rǎng bǎo yù xié yì)

An agreement entered into under soil legislation for the purpose of conserving the ecological integrity of soil. [Drafting Legislation for Sustainable Soils: A Guide, p. 92 (2004)]

Soil Degradation (土壤退化) (tǔ rǎng tuì huà)

Includes aspects of physical, chemical and/or biological deterioration. Examples are loss of organic matter, decline in soil fertility, decline in structural condition, soil erosion, adverse changes in salinity, acidity or alkalinity and the effects of toxic chemicals, pollutants or excessive flooding. [Drafting Legislation for Sustainable Soils: A Guide, p. 93 (2004)]

Soil Ecological Community (土壤生态群落) (tǔ rǎng shēng tài qún luò)

A generic term which means an area of land which has been identified by mapping, and according to specified soil ecological criteria, that interacts as an identifiable functional or spatial unit. It is synonymous with a "soil landscape." Other terms that may also be used interchangeably with the concept of a "soil ecological community," include "soil quality," "soil type," "soil ecosystems," "soil resources." [Drafting Legislation for Sustainable Soils: A Guide, p. 93 (2004)]

Soil Environment (土壤环境) (tǔ rǎng huán jìng)

The natural dynamic system of unconsolidated mineral and organic material at the Earth's surface, which interacts with the living community of earth and the natural, human-made and social surroundings of that community. [Drafting Legislation for Sustainable Soils: A Guide, p. 93 (2004)]

Soil Environmental Impact Assessment (土壤环境影响评价) (tǔ rǎng huán jìng yǐng xiǎng píng jià)

A report evaluating the likely impact of an activity on the ecological integrity of soil. [Drafting Legislation for Sustainable Soils: A Guide, p. 93 (2004)]

Soil Gas (土壤气体) (tǔ rǎng qì tǐ)

The gas present in soil which may contain radon. [US EPA Glossary of Indoor Air Quality Terms (2011)]

Soil Horizons (土层) (tǔ céng)

A layer of soil having distinct characteristics produced by soil forming processes. [Urban Soil Primer, p. 72 (2005)]

Soil Landscape (土壤景观) (tǔ rǎng jǐng guān)

An area of land that has recognisable and specifiable topography that is capable of being presented on maps and of being described by concise statements. Thus, a soil landscape has a characteristic landform with one or more soil taxonomic units occurring in a defined way. It is often associated with the physiographic features of the landscape and is similar to a soil association, but in a soil landscape the landform pattern is specifically described. The soil landscape may be named according to the soil taxonomic units present, the dominant unit, or be given a geographic name based on a locality where it is well developed. [Drafting Legislation for Sustainable Soils: A Guide, p. 93 (2004)]

Soil Profile (土壤剖面) (tǔ rǎng pōu miàn)

A vertical section of the soil extending through all its horizons and into the parent material. [Urban Soil Primer, p. 73 (2005)]

Soil Quality (土壤质量) (tǔ rǎng zhí liàng)

"Soil quality" means the capacity of a specific type of soil to function, within natural or fabricated ecosystems or land use boundaries, to sustain plant and animal productivity, maintain or enhance water and air quality and support human health and habitation. [Drafting Legislation for Sustainable Soils: A Guide, p. 93 (2004)]

Soil Reaction (土壤反应) (tǔ rǎng fǎn yìng)

A measure of acidity or alkalinity of a soil, expressed in pH values. A soil that tests to pH 7.0 is described as precisely neutral in reaction because it is neither acid nor alkaline. [Urban Soil Primer, p. 73 (2005)]

Soil Series (土壤系列) (tǔ rǎng xì liè)

A group of soils that have profiles that are almost alike, except for differences in texture of the surface layer. All of the soils of a series have horizons that are similar in composition, thickness, and arrangement. [Urban Soil Primer, p. 73 (2005)]

Soil Structure (土壤结构) (tǔ rǎng jié gòu)

The arrangement of primary soil particles into compound particles or aggregates. The principal forms of soil are platy, prismatic, columnar, blocky, and granular. Structureless soils are either grained or massive. [Urban Soil Primer, p. 74 (2005)]

Soil Texture (土壤质地) (tǔ rǎng zhí dì)

The relative proportions of sand, silt, and clay particles in a mass of soil. The basic textural classes, in order of increasing proportion of fine particles, are sand, loamy sand, sandy loam, loam, silt loam, silt, sandy clay loam, clay loam, silty clay loam, sandy clay, silty clay, and clay. The sand, loamy sand, and sandy loam classes may be further divided by specifying "coarse," "fine," or "very fine." [Urban Soil Primer, p. 74 (2005)]

Soil-forming Factors (土壤形成因素) (tǔ rǎng xíng chéng yīn sù)

Five factors responsible for the formation of the soil from the unconsidated parent material. The factors are time, climate, parent material, living organisms (including humans), and relief. [Urban Soil Primer, p. 74 (2005)]

Soil-gas-retarder (土壤气体缓凝剂) (tǔ rǎng qì tǐ huǎn níng jì)

A continuous membrane or other comparable material used to retard the flow of soil gases into a building. [US EPA Glossary of Indoor Air Quality Terms (2011)]

Solar Radiation (太阳辐射) (tài yáng fú shè)

Radiation emitted by the sun. It is also referred to as short-wave radiation. Solar radiation has a distinctive range of wavelengths (spectrum) determined by the temperature of the sun. [US EPA Glossary of Climate Change Terms (2011)]

Source (源) (yuán)

Any process or activity which releases a greenhouse gas, an aerosol or a precursor of a greenhouse gas into the atmosphere. [United Nations Framework Convention on Climate Change, Art. 1, No. 9 (1992)]

Sources of Indoor Air Pollutants (源 (室内空气污染)) (yuán (shì nèi kōng qì wū rǎn))

Indoor air pollutants can originate within the building or be drawn in from outdoors. Common sources include people, room furnishings such as carpeting, photocopiers, art supplies, etc. [US EPA Glossary of Indoor Air Quality Terms (2011)]

Sources of International Environmental Law (国际环境法渊源) (guó jì huán jìng fǎ yuān yuán)

The treaties and multilateral environmental agreements, together with the general principles of environmental law as found in the Rio Declaration on Environment and Development, and customary international law and environmental law, as evaluated by scholarly and professional legal commentaries as publicists of public international law, are the sources of international environmental law.

Sovereign Right of States to Exploit their Natural Resources (国家开发其自然资源的主权权利) (guó jiā kāi fā qí zì rán zī yuán de zhǔ quán quán lì)

States have the sovereign right to exploit their natural resources pursuant to their environmental policies and in accordance with their duty to protect and preserve the marine environment. [Law of the Sea Convention, Art. 193 (1982)]

Special Exception Permit (特别例外许可证) (tè bié lì wài xǔ kě zhèng)

See **Special Use Permit**.

Special Use Permit (特殊用途许可证) (tè shū yòng tú xǔ kě zhèng)

Special uses are allowed in zoning districts, but only upon the issuance of a special use permit subject to conditions designed to protect surrounding properties and the neighborhood from the negative impacts of the permitted use. Also called conditional use permit, special exception permit, and special permit. [Nolon, Well Grounded, p. 455 (ELI 2001)]

Species (种) (zhǒng)

A reproductively isolated set of interbreeding populations of living organisms.

Specific Legal Regime of the Exclusive Economic Zone (专属经济区的特定法律制度) (zhuān shǔ jīng jì qū de tè dìng fǎ lǜ zhì dù)

The exclusive economic zone is an area beyond and adjacent to the territorial sea, subject to the specific legal regime established in this Part, under which the rights and jurisdiction of the coastal State and the rights and freedoms of other States are governed by the relevant provisions of this Convention. [Law of the Sea Convention, Art. 55 (1982)]

Specification (规格) (guī gé)

The document that prescribes the requirements with which the product or service must conform. [Von Zharen, ISO 14000, pp. 204 (1996)]

Spot Zoning (小块土地分区) (xiǎo kuǎi tǔ dì fēn qū)

The rezoning of a single parcel or a small area to benefit one or more property owners rather than to carry out an objective of the comprehensive plan. [Nolon, Well Grounded, p. 455 (ELI 2001)]

Stack Effect (烟囱效应) (yān cōng xiào yìng)

The overall upward movement of air inside a building that results from heated air rising and escaping through openings in the building super structure, thus causing an indoor pressure level lower than that in the soil gas beneath or surrounding the building foundation. [US EPA Glossary of Indoor Air Quality Terms (2011)]

Stakeholders (利益相关者) (lì yì xiāng guān zhě)

Those groups and organizations having an interest or stake in a company's EMS program (e.g., regulators, shareholders, customers, suppliers, special interest groups, residents, competitors, investors, bankers, media, lawyers, insurance companies, trade groups, unions, ecosystems, cultural heritage, and geology). [Von Zharen, ISO 14000, pp. 204–5 (1996)]

State of Export (出口国) (chū kǒu guó)

A Party from which a transboundary movement of hazardous wastes or other wastes is planned to be initiated or is initiated. [Basel Convention, Art. 2 (1989)]

State of Import (进口国) (jìn kǒu guó)

A Party to which a transboundary movement of hazardous wastes or other wastes is planned or takes place for the purpose of disposal therein or for the purpose of loading prior to disposal in an area not under the national jurisdiction of any State. [Basel Convention, Art. 2 (1989)]

State of Transit (过境国) (guò jìng guó)

Any State, other than the State of export or import, through which a movement of hazardous wastes or other wastes is planned or takes place. [Basel Convention, Art. 2 (1989)]

State Soil Strategy (国家土壤战略) (guó jiā tǔ rǎng zhàn lüè)

The strategy prepared under national soil legislation to set out how the objectives of soil legislation will be achieved. [Drafting Legislation for Sustainable Soils: A Guide, p. 93 (2004)]

States Concerned (有关国家) (yǒu guān guó jiā)

Parties which are States of export or import, or transit States, whether or not Parties. [Basel Convention, Art. 2 (1989)]

Static Pressure (静压) (jìng yā)

Condition that exists when an equal amount of air is supplied to and exhausted from a space. At static pressure, equilibrium has been reached. [US EPA Glossary of Indoor Air Quality Terms (2011)]

Statute of Limitations (时效) (shí xiào)

A law that requires that an aggrieved party file a legal action in a quasi-judicial or judicial forum within a specified period or lose the right to file the action. Regarding many land use determinations, the period begins from the date the determination is filed with the municipal clerk. [Nolon, Well Grounded, p. 455 (ELI 2001)]

Sterilizer (灭菌) (miè jùn)

One of three groups of antimicrobials registered by EPA for public health uses. EPA considers an antimicrobial to be a sterilizer when it destroys or eliminates all forms of bacteria, fungi, viruses, and their spores. Because spores are considered the most difficult form of a microorganism to destroy, EPA considers the term sporicide to be synonymous with "sterilizer." [US EPA Glossary of Indoor Air Quality Terms (2011)]

Stern Report (斯特恩报告) (sī tè ēn bào gào)

The Stern Review on the Economics of Climate Change is a 700-page report written in 2006 by the British economist Sir Nicholas Stern for the UK Government, which discusses the effect of climate change and global warming on the world economy. [UNEP Guide to Climate Neutrality (2008)]

Stockholm Convention on Persistent Organic Pollutants (关于持久性有机污染物的斯德哥尔摩公约) (guān yú chí jiǔ xìng yǒu jī wū rǎn wù de sī dé gē ěr mó gōng yuē)

This Convention entered into force May17, 2004. The Convention aims to restrict or eliminate the production of persistent organic pollutants. [Stockholm Convention (2001)]

Storage (贮藏) (chú cáng)

Holding products, materials, substances, goods, wastes or other objects on a temporary or permanent basis.

Strategies (策略) (cè lüè)

A set of actions to be undertaken to accomplish each objective contained in a comprehensive plan. To obtain the objective of "50 units of affordable housing," the plan may include strategies: (1) Form a housing trust fund, and (2) Allow for accessory apartments in residential units. [Nolon, Well Grounded, p. 455 (ELI 2001)]

Stratosphere (平流层) (píng liú céng)

Region of the atmosphere between the troposphere and mesosphere, having a lower boundary of approximately 8 km at the poles to 15 km at the equator and an upper boundary of approximately 50 km. Depending upon latitude and season, the temperature in the lower stratosphere can increase, be isothermal, or even decrease with altitude, but the temperature in the upper stratosphere generally increases with height due to absorption of solar radiation by ozone. [US EPA Glossary of Climate Change Terms (2011)]

Streamflow (流量) (liú liàng)

The volume of water that moves over a designated point over a fixed period of time. It is often expressed as cubic feet per second (ft^3/sec). [US EPA Glossary of Climate Change Terms (2011)]

Strict Liability (严格责任) (yán gé zé rèn)

The absolute liability for making both compensation and reparations for harm that that results for a party when that harm results from the use of an inherently dangerous object or materials, without any requirement to show that the party was at fault.

Strict Nature Reserve: Protected Area Managed Mainly for Science (严格自然保护区：主要为科研而管理的保护区) (yán gé zì rán bǎo hù qū: zhǔ yào wéi kē yán ér guǎn lǐ de bǎo hù qū)

Area of land and/or sea possessing some outstanding or representative ecosystems, geological or physiological features and/or species, available primarily for scientific research and/or environmental monitoring. [Guidelines for Protected Area Management Categories (1994)]

Studies, Research Programmes and Exchange of Information and Data (学习、研究计划及信息、数据交换) (xué xí, yán jiū jì huà jí xìng xī shù jù jiāo huàn)

States shall cooperate, directly or through competent international organizations, for the purpose of promoting studies, undertaking programmes of scientific research and encouraging the exchange of information and data acquired about pollution of the marine environment. They shall endeavour to participate actively in regional and global programmes to acquire knowledge for the assessment of the nature and extent of pollution, exposure to it, and its pathways, risks and remedies. [Law of the Sea Convention, Art. 200 (1982)]

Subcontractor (分包商) (fēn bāo shāng)

An organization that provides a product to the supplier. [Von Zharen, ISO 14000, p. 205 (1996)]

Subdivision (再分) (zài fēn)

The subdivision of land involves the legal division of a parcel into a number of lots for the purpose of development and sale. The subdivision and development of individual parcels must conform to the provisions of local zoning which contain use and dimensional requirements for land development. [Nolon, Well Grounded, p. 455 (ELI 2001)]

Subdivision Plat (小块土地) (xiǎo kuài tǔ dì)

See **Plat**.

Subject of International Environmental Law (国际环境法主体) (guó jì huán jìng fǎ zhǔ tǐ)

A sovereign State is the primary subject of international environmental law, although some agreements give recognition to and provide for individuals as environmental experts or for species when endangered, or for political subdivisions of States or governmental agencies within States.

Sub-membrane Depressurization System (子膜卸压系统) (zǐ mó xiè yā xì tǒng)

A system designed to achieve lower sub-membrane air pressure relative to crawlspace air pressure by use of a fan-powered vent drawing air from under the soil-gas-retarder membrane. [US EPA Glossary of Indoor Air Quality Terms (2011)]

Subsidiary Bodies (附属机构) (fù shǔ jī gòu)

As authorized by an international agreement, States may create new inter-government bodies to carry out the mandates of the agreement and its conference of the parties, known generically as subsidiary bodies.

Subsidiary Body for Implementation (UNFCCC) (附属履行机构) (fù shǔ lǚ xíng jī gòu)

A subsidiary body for implementation is hereby established to assist the Conference of the Parties in the assessment and review of the effective implementation of the Convention. This body shall be open to participation by all Parties and comprise government representatives who are experts on matters related to climate change. It shall report regularly to the Conference of the Parties on all aspects of its work. [United Nations Framework Convention on Climate Change, Art. 10 (1992)]

Subsidiary Body for Scientific and Technological Advice (UNFCCC) (附属科技咨询机构) (fù shǔ kē jì zī xún jī gòu)

A subsidiary body for scientific and technological advice is hereby established to provide the Conference of the Parties and, as appropriate, its other subsidiary bodies with timely information and advice on scientific and technological matters relating to the Convention. This body shall be open to participation by all Parties and shall be multidisciplinary. It shall report regularly to the Conference of the Parties on all aspects of its work. [United Nations Framework Convention on Climate Change, Art. 9 (1992)]

Sub-slab Depressurization System (Active) (子板卸压系统 (主动)) (zǐ bǎn xiè yā xì tǒng (zhǔ dòng))

A system designed to achieve lower sub-slab air pressure relative to indoor air pressure by use of a fan-powered vent drawing air from beneath the slab. [US EPA Glossary of Indoor Air Quality Terms (2011)]

Sub-slab Depressurization System (Passive) (子板卸压系统 (被动)) (zǐ bǎn xiè yā xì tǒng (bèi dòng))

A system designed to achieve lower sub-slab air pressure relative to indoor air pressure by use of a vent pipe routed through the conditioned space of a building and connecting the sub-slab area with outdoor air, thereby relying solely on the convective flow of air upward in the vent to draw air from beneath the slab. [US EPA Glossary of Indoor Air Quality Terms (2011)]

Subsoil (下层土) (xià céng tǔ)

The soils and minerals found beneath the layer of topsoil.

Sulfate Aerosols (硫酸盐气溶胶) (liú suān yán qì róng jiāo)

Particulate matter that consists of compounds of sulfur formed by the interaction of sulfur dioxide and sulfur trioxide with other compounds in the atmosphere. Sulfate aerosols are injected into the atmosphere from the combustion of fossil fuels and the eruption of volcanoes like Mt. Pinatubo. Recent theory suggests that sulfate aerosols may lower the Earth's temperature by reflecting away solar radiation (negative radiative forcing). General Circulation Models which incorporate the effects of sulfate aerosols more accurately predict global temperature variations. [US EPA Glossary of Climate Change Terms (2011)]

Sulfur Hexafluoride (SF^6) (六氟化硫) (liù fú huà liú)

A colorless gas soluble in alcohol and ether, slightly soluble in water. A very powerful greenhouse gas used primarily in electrical transmission and distribution systems and as a dielectric in electronics. The global warming potential [GWP] of SF_6 is 22 200. This GWP is from the IPCC's Third Assessment Report (TAR). [US EPA Glossary of Climate Change Terms (2011)]

Supplier (供应商) (gōng yìng shāng)

An organization that provides a product to the customer. [Von Zharen, ISO 14000, p. 205 (1996)]

Surface Drainage (地表排水) (dì biǎo pái shuǐ)

Runoff, or surface flow of water from an area. [Urban Soil Primer, p. 71 (2005)]

Survey (调查) (diào chá)

An examination for some specific purpose: to inspect or consider carefully; to review in detail. [Von Zharen, ISO 14000, p. 205 (1996)]

Sustainable Development (可持续发展) (kě chí xù fā zhǎn)

Policies and measures to protect the climate system against human-induced change should be appropriate for the specific conditions of each Party and should be integrated with national development programmes, taking into account that economic development is essential for adopting measures to address climate change. [United Nations Framework Convention on Climate Change, Art. 3 (1992)]

Sustainable Development and Agenda 21 (可持续发展与21世纪议程) (ké chí xù fā zhǎn yǔ èr shí yī shì jì yì chéng)

The Rio de Janeiro United Nations Conference on Environment and Development in 1992 redefined socio-economic development as sustainable development, with the recommended criteria for sustainable development set forth in the written report of the Conference, which was agreed to by all States by consensus with the name Agenda 21.

Sustainable Use (可持续利用) (kě chí xù lì yòng)

The use of components of biological diversity in a way and at a rate that does not lead to the long-term decline of biological diversity, thereby maintaining its potential to meet the needs and aspirations of present and future generations. [Convention on Biological Diversity, Art. 2 (1992)]

Sustainable Use and Conservation (可持续利用和保育) (kě chí xù lì yòng hé bǎo yù)

The use and harvesting of renewable natural resources of plants and animals, soils, and water, so as not to deplete the specific resource and to allow its sustained reproduction for future yields.

Sustainable Use of Soil (土壤的可持续利用) (tǔ rǎng de kě chí xù lì yòng)

The use of soils in a manner that preserves the balance between the processes of soil formation and soil degradation, while maintaining the ecological functions and needs of soil. In this context, "the use of soil" means "the role of soil in the conservation of biodiversity and the maintenance of human life." [Drafting Legislation for Sustainable Soils: A Guide, p. 93 (2004)]

System Boundary (系统边界) (xì tǒng biān jiè)

Interface between the product or service system being studied and its environment or other systems. [Von Zharen, ISO 14000, p. 205 (1996)]

T

Taking (征收) (zhēng shōu)

See **Regulatory Taking**.

Tax Policy on Energy Resource Saving (节约能源资源税收政策) (jié yuē néng yuán zī yuán shuì shōu zhèng cè)

The laws and regulations for taxation that establish norms to reduce taxes when income-producing activities are operated to save energy as an inducement to undertake energy-saving measures.

Technical Expert (技术专家) (jì shù zhuān jiā)

Individual who provides specific knowledge or expertise to the audit team, but who does not participate as an auditor. [Von Zharen, ISO 14000, p. 205 (1996)]

Technical Regulation (技术法规) (jì shù fǎ guī)

[A] document which lays down product characteristics or their related processes and production methods, including the applicable administrative provisions, with which compliance is mandatory. It may also include or deal exclusively with terminology, symbols, packaging, marking or labelling requirements as they apply to a product, process or production method. [WTO, Agreement on Technical Barriers to Trade (1995)]

Technology (技术) (jì shù)

(1) The method and application of science to achieve a practical purpose.
(2) Includes biotechnology. [Convention on Biological Diversity (1992)]

Tennessee Valley Authority v. Hill Case
(田纳西流域管理局诉希尔案) (tián nà xī liú yù guǎn lǐ jú sù xī ér àn)

A 1978 case in which the US Supreme Court decided that the Endangered Species Act prohibited the completion of a dam due to the presence of a population of Snail Darters (small, endangered fish). [TVA v. Hill (1978)]

Testing (试验) (shì yàn)

A means of determining an item's capability to meet specified requirements by subjecting them to a set of physical, chemical, environmental, or operating actions and conditions. [Von Zharen, ISO 14000, p. 205 (1996)]

Thermohaline Circulation (温盐环流) (wēn yán huán liú)

Large-scale density-driven circulation in the ocean, caused by differences in temperature and salinity. In the North Atlantic the thermohaline circulation consists of warm surface water flowing northward and cold deep water flowing southward, resulting in a net poleward transport of heat. The surface water sinks in highly restricted sinking regions located in high latitudes. [US EPA Glossary of Climate Change Terms (2011)]

Third Party (第三方) (dì sān fāng)

Person or legal entity recognized as being independent of the parties involved in the sale of a product. The practitioner, or its agent, is a third party. Suppliers or producers are the first party and consumers the second party. [Von Zharen, ISO 14000, p. 205 (1996)]

Third Party Transfer (第三方转让) (dì sān fāng zhuǎn ràng)

Transfer of material, knowledge and/or products arising from access to a party other than the user as identified in the PIC [Prior Informed Consent] and MAT [Mutually Agreed Terms] agreements. [UNU IAS Pocket Guide, p. 16 (2007)]

Three Mechanisms Of The Kyoto Protocol
(京都议定书三机制) (jīng dū yì dìng shū sān jī zhì)

Emissions trading, the Clean Development Mechanism (CDM), and Joint Implementation (JI). [UNFCCC, The Mechanisms Under the Kyoto Protocol (1997)]

Three Mile Island Nuclear Power Plant Spill (United States)
(三哩岛核电站泄漏事故) (美国) (sān lǐ dǎo hé diàn zhàn xiè lòu shì gù (měi guó))

A nuclear meltdown on March 28, 1979, at Three Mile Island power plant in the state of Pennsylvania, United States, resulting in the release of radioactive material.

TLVs (阈值) (yù zhí)

Threshold Limit Values (guidelines recommended by the American Conference of Governmental Industrial Hygienists). [US EPA Glossary of Indoor Air Quality Terms (2011)]

Tolerable Limits (允许限度) (yǔn xǔ xiàn dù)

Governmentally permissible residue levels for chemicals in the food chain or in other regulated systems or substances.

Topographic Maps (地形图) (dì xíng tú)

Maps that show terrain, ridges, waterways, contours, elevations, and geographic locations. [Urban Soil Primer, p. 74 (2005)]

Town Board (镇委会) (zhèng wěi huì)

See **Local Legislature**.

Toxic Substances Control Act (TSCA, United States) (有毒物质控制法 (美国)) (yǒu dú wù zhí kòng zhì fǎ (měi guó))

[A federal statute that] was designed to slow down the production of toxic substances, and prevent those produced from presenting an unreasonable risk of injury to one's health or to the environment. A knowing or wilful violation of this act is punishable by up to one year imprisonment. [UNEP Judicial Handbook, p. 81 (2005)] Enacted in 1976 to address the production, importation, use, and disposal of specific chemicals. [US EPA, TSCA (2012)]

Toxicity (毒性) (dú xìng)

(1) The degree to which something is toxic. (2) The danger of harm posed by a substance for a living organism, ultimately harm that can result in death.

Trace Elements (微量元素) (wēi liàng yuán sù)

Chemical elements which are available in extremely small amounts that are essential to plant growth. [Urban Soil Primer, p. 74 (2005)]

Trace Gases (微量气体) (wēi liàng qì tǐ)

(1) Any one of the less common gases found in the Earth's atmosphere. Nitrogen, oxygen, and argon make up more than 99 percent of the Earth's atmosphere. Other gases, such as carbon dioxide, water vapor, methane, oxides of nitrogen, ozone, and ammonia, are considered trace gases. Although relatively unimportant in terms of their absolute volume, they have significant effects on the Earth's weather and climate. [US EPA Glossary of Climate Change Terms (2011)] (2) Compounds, such as sulfur hexafluoride, which are used to identify suspected pollutant pathways and to quantify ventilation rates. Trace gases may be detected qualitatively by their odor or quantitatively by air monitoring equipment. [US EPA Glossary of Indoor Air Quality Terms (2011)]

Traditional Biomass (传统的生物质能) (chuán tǒng de shēng wù zhì néng)

(1) The sum or characteristic components of uncultivated or wild and indigenous plant organisms endemic in a geographic place. (2) Wood and other fuels derived from plants that have traditionally been used to create heat and energy.

Traditional Knowledge (传统知识) (chuán tǒng zhī shì)

The knowledge, innovations and practices of local and indigenous communities relevant to conservation and sustainable use of biological diversity. [UNU IAS Pocket Guide, p. 15 (2007)]

Traffic in Transit (中转交通) (zhōng zhuǎn jiāo tōng)

The transit of persons, baggage, goods and means of transport across the territory of one or more transit States, when the passage across such territory, with or without trans-shipment, warehousing, breaking bulk or change in the mode of transport, is only a portion of a complete journey which begins or terminates within the territory of the land-locked State. [Law of the Sea Convention, Art. 124 (1982)]

Trail Smelter Arbitration (United States and Canada) (特雷尔冶炼厂仲裁案) (tè léi ěr yě liàn chǎng zhòng cái àn)

A 1905 arbitration between the United States and Canada to resolve a dispute over environmental damage from a smelter in Trail, British Columbia. [Trail Smelter Case (1905)]

Transboundary Exposure (跨界暴露) (kuà jiè bào lù)

Exposure of members of the public in one State due to radioactive material released via accidents, discharges or waste disposal in another State. [IAEA Safety Glossary, p. 211 (2007)]

Transboundary Movement (越境转移) (yuè jìng zhuǎn yí)

(1) Any movement of hazardous wastes or other wastes from an area under the national jurisdiction of one State to or through an area under the national jurisdiction of another State or to or through an area not under the national jurisdiction of any State, provided at least two States are involved in the movement. [Basel Convention Art II § 3 (1989)] (2) Any movement of radioactive material from one State to or through another or any shipment of spent fuel or of radioactive waste from a State of origin to a State of destination. [IAEA Safety Glossary, p. 211 (2007)]

Transfer of Development Rights (开发权转让) (kāi fā quán zhuǎn ràng)

Provisions in a zoning law that allow for the purchase of the right to develop land located in a sending area and the transfer of those rights to land located in a receiving area. [Nolon, Well Grounded, p. 456 (ELI 2001)]

Transient Population Groups (暂居人群) (zàn jū rén qún)

Those members of the public who are residing for a short period of time (days to weeks) in a location (such as a camping ground) that can be identified in advance. This does not include members of the public who may be travelling through an area. [IAEA Safety Glossary, p. 211 (2007)]

Transit State (过境国) (guò jìng guó)

A State, with or without a sea-coast, situated between a land-locked State and the sea, through whose territory traffic in transit passes. [Law of the Sea Convention, Art. 124 (1982)]

Treaty (条约) (tiáo yuē)

An international agreement concluded between States in written form and governed by international law, whether embodied in a single instrument or in two or more related instruments and whatever its particular designation. [UNEP Judicial Handbook, p. 39 (2005)]

Troposphere (对流层) (duì liú céng)

The lowest part of the atmosphere from the surface to about 10 km in altitude in mid-latitudes (ranging from 9 km in high latitudes to 16 km in the tropics on average) where clouds and "weather" phenomena occur. In the troposphere temperatures generally decrease with height. [US EPA Glossary of Climate Change Terms (2011)]

Trust Fund for the Least Developed Countries (最不发达国家信托基金) (zuì bù fā dá guó jiā xìn tuō jī jīn)

A voluntary trust fund established under the UNFCCC to address the special needs of the 48 Least Developed Countries (LDCs) that are especially vulnerable to the adverse impacts of climate change. To date, the fund has 19 donors: Australia, Austria, Canada, Denmark, Finland, France, Germany, Ireland, Italy, Japan, Luxembourg, Netherlands, New Zealand, Norway, Portugal, Spain, Sweden, Switzerland, and United Kingdom. [GEF-Administered Trust Funds (2010)]

Trusteeship Council (托管理事会) (tuō guǎn lǐ shì huì)

The Charter established the Trusteeship Council as one of the main organs of the United Nations and assigned to it the task of supervising the administration of Trust Territories placed under the Trusteeship System. Major goals of the System were to promote the advancement of the inhabitants of Trust Territories and their progressive development towards self-government or independence. The Trusteeship Council is made up of the five permanent members of the Security Council: China, France, Russian Federation, United Kingdom and United States. The aims of the Trusteeship System have been fulfilled to such an extent that all Trust Territories have attained self-government or independence, either as separate States or by joining neighbouring independent countries. The Trusteeship Council suspended operation on 1 November 1994, with the independence of Palau, the last remaining United Nations trust territory, on 1 October 1994. [United Nations, Trusteeship Council (2012)]

Tuna Case (European Community v. United States) (DS29/R, 1994) (金枪鱼案 (欧共体诉美国) (DS29/R, 1994)) (jīn qiāng yú àn (ōu gòng tǐ sù měi guó) (DS29/R, 1994))

A 1994 GATT Panel report, requested by the European Economic Community, on the restrictions maintained by the United States on the importation of certain tuna products. [Restrictions on Imports of Tuna (1994)]

TVOCs (总挥发性有机化合物) (zǒng huī fā xìng yǒu jī huà hé wù)

Total volatile organic compounds. [US EPA Glossary of Indoor Air Quality Terms (2011)]

Type I Action (类型I活动) (lèi xíng yī huó dòng)

Under New York's SEQRA, an action that is more likely to have an adverse impact on the environment than an unlisted action. [Nolon, Well Grounded, p. 456 (ELI 2001)]

Type II Action (类型II活动) (lèi xíng èr huó dòng)

An action that is not subject to environmental review under New York's SEQRA.... [Nolon, Well Grounded, p. 456 (ELI 2001)]

U

UK Environmental Law (英国环境法) (yīng guó huán jìng fǎ)

A body of law in the United Kingdom regarding generally air, water, and land.

UK London Smog Event (英国伦敦烟雾事件) (yīng guó lún dūn yān wù shì jiàn)

A weather event in December 1952 due to an unusually long cold period and a temperature inversion. The result was dense smog that lasted for 5 days and caused many deaths.

Ultraviolet Radiation (UV) (紫外线) (zǐ wài xiàn)

The energy range just beyond the violet end of the visible spectrum. Although ultraviolet radiation constitutes only about 5 percent of the total energy emitted from the sun, it is the major energy source for the stratosphere and mesosphere, playing a dominant role in both energy balance and chemical composition. Most ultraviolet radiation is blocked by Earth's atmosphere, but some solar ultraviolet penetrates and aids in plant photosynthesis and helps produce vitamin D in humans. Too much ultraviolet radiation can burn the skin, cause skin cancer and cataracts, and damage vegetation. [US EPA Glossary of Climate Change Terms (2011)]

Unauthorized Broadcasting From The High Seas (从公海未经许可的广播) (cóng gōng hǎi wèi jīng xǔ kě de guǎng bò)

1. All States shall cooperate in the suppression of unauthorized broadcasting from the high seas. 2. For the purposes of this Convention, "unauthorized broadcasting" means the transmission of sound radio or television broadcasts from a ship or installation on the high seas intended for reception by the general public contrary to international regulations, but excluding the transmission of distress calls. 3. Any person engaged in unauthorized broadcasting may be prosecuted before the court of: (a)

the flag State of the ship; (b) the State of registry of the installation; (c) the State of which the person is a national; (d) any State where the transmissions can be received; or (e) any State where authorized radio communication is suffering interference. 4. On the high seas, a State having jurisdiction in accordance with paragraph 3 may, in conformity with article 110, arrest any person or ship engaged in unauthorized broadcasting and seize the broadcasting apparatus. [Law of the Sea Convention, Art. 109 (1982)]

Unharmful Materials (无害物质) (wú hài wù zhí)

[Ocean dumping] Materials where the primary concern is physical impact. Article 4 of the 1996 Protocol to the London Dumping Convention states that Contracting Parties "shall prohibit the dumping of any wastes or other matter with the exception of those listed in Annex 1." Annex I includes "7. Bulky items primarily comprising iron, steel, concrete and similar unharmful materials for which the concern is physical impact and limited to those circumstances, where such wastes are generated at locations, such as small islands with isolated communities, having no practicable access to disposal options other than dumping." [1972 Convention on the Prevention of Marine Pollution by Dumping of Wastes and Other Matter, 1996 Protocol, Annex I, Item 7]

Unit Ventilator (单元通风机) (dān yuán tōng fēng jī)

A fan-coil unit package device for applications in which the use of outdoor- and return-air mixing is intended to satisfy tempering requirements and ventilation needs. [US EPA Glossary of Indoor Air Quality Terms (2011)]

United Nations Convention on the Law of the Sea (UNCLOS) (联合国海洋法公约) (lián hé guó hǎi yáng fǎ gōng yuē)

A treaty adopted on December 10, 1982, to protect fishery resources and recognize the sovereign rights of the coastal States to explore and exploit their resources.

United Nations Convention to Combat Desertification (联合国防止荒漠化公约) (lián hé guó fáng zhǐ huāng mò huà gōng yuē)

A treaty adopted on June 17, 1994, to maintain and restore land and soil productivity.

United Nations Development Program (UNDP) (联合国开发计划署) (lián hé guó kāi fā jì huà shǔ)

The United Nations' global development network, an organization has been advocating for change and connecting countries to knowledge, experience and resources to help people build a better life. [UNDP (2012)]

United Nations Educational, Scientific and Cultural Organization (UNESCO) (联合国教科文组织) (lián hé guó jiào kē wén zǔ zhī)

An organization working to achieve global visions of sustainable development encompassing observance of human rights, mutual respect and the alleviation of poverty. [UNESCO (2012)]

United Nations Environment Programme (UNEP) (联合国环境规划署) (lián hé guó huán jìng guī huà shǔ)

An organization working to provide leadership and encourage partnership in caring for the environment by inspiring, informing, and enabling nations and peoples to improve their quality of life without compromising that of future generations. [UNEP (2012)]

United Nations Food and Agriculture Organization (FAO) (联合国粮农组织) (lián hé guó liáng nóng zǔ zhī)

A United Nations organization working to achieve food security for all; to make sure people have regular access to enough high-quality food to lead active, healthy lives; and to raise levels of nutrition, improve agricultural productivity, better the lives of rural populations and contribute to the growth of the world economy. [FAO (2012)]

United Nations Framework Convention on Climate Change (UNFCCC) (联合国气候变化框架公约) (lián hé guó qì hòu biàn huà kuāng jià gōng yuē)

A treaty to combat climate change, opened for signature on May 9, 1992. (1) The United Nations Framework Convention on Climate Change (UNFCCC) is the first international climate treaty. It came into force in 1994 and has since been ratified by 189 countries including the United States. More recently, a number of nations have approved an addition to the treaty: the Kyoto Protocol, which has more powerful (and legally binding) measures. [UNEP Guide to Climate Neutrality (2008)] (2) The Convention on Climate Change sets an overall framework for intergovernmental efforts to tackle the challenge posed by climate change. It recognizes that the climate system is a shared resource whose stability can be affected by industrial and other emissions of carbon dioxide and other greenhouse gases. The Convention enjoys near universal membership, with 189 countries having ratified. Under the Convention, governments gather and share information on greenhouse gas emissions, national policies and best practices; launch national strategies for addressing greenhouse gas emissions and adapting to expected impacts, including the provision of financial and technological support to developing countries; cooperate in preparing for adaptation to the impacts of climate change. The Convention entered into force on 21 March 1994. [US EPA Glossary of Climate Change Terms (2011)]

United Nations International Convention on the Law of the Non-navigational Uses of International Watercourses (联合国国际水道非航利用法公约) (lián hé guó guó jì shuǐ dào fēi háng lì yòng fǎ gōng yuē)

Adopted by the United Nations on May 21, 1997, this document provides international guidelines for the use of both surface and ground waters that cross international boundaries.

Unlisted Actions (未编入册的活动) (wèi biān rù cè de huó dòng)

These are all of the actions that are not listed as Type I or Type II actions for the purposes of New York's SEQRA. These actions are subject to review by the lead agency to determine whether they may cause significant adverse environmental impacts. [Nolon, Well Grounded, p. 456 (ELI 2001)] See **Type I Action** and **Type II Action**.

Upwelling (涌升流) (yǒng shēng liú)

The rising of colder, denser water from deep in the ocean or other body of water to the surface, often as a result of wind-driven surface currents. Upwelling can occur anywhere, but is most common in coastal areas and along the equator. Because nutrients accumulate in deeper ocean water through the remains and waste of marine life, upwelling fertilizes surface waters and increases productivity. [NASA Earth Observatory (2012)]

Use District (用途分区) (yòng tú fēn qū)

See **Zoning District**.

Use of Water Resources of International Rivers, the Helsinki Rules (国际河流水资源利用, 赫尔辛基规则) (guó jì hé liú shuǐ zī yuán lì yòng, hè'ěr xīn jī guī zé)

International guidelines adopted in August 1966 regarding the use of rivers and related groundwater that crosses national borders.

Use Variance (用途变动) (yòng tú biàn dòng)

A variance that allows a landowner to put his land to a use that is not permitted under the zoning law, for example, if a parcel of land is zoned for single-family residential use and the owner wishes to operate a retail business, the owner must apply to the zoning board of appeals for a use variance. A use variance may be granted only in cases of unnecessary hardship. To prove unnecessary hardship, the owner must establish that the requested variance meets four statutorily prescribed conditions. [Nolon, Well Grounded, p. 456 (ELI 2001)]

V

Vapor Recovery System (气体回收系统) (qì tǐ huí shōu xì tǒng)

An anti-pollution system designed to capture gasoline vapors that would otherwise escape into the atmosphere from hot vehicle engines and fuel tanks. [US EPA Glossary of Mobile Source Emissions Terms (2012)]

Variable Air Volume System (VAV) (变量空气系统) (biàn liàng kōng qì xì tǒng)

Air handling system that conditions the air to constant temperature and varies the outside airflow to ensure thermal comfort. [US EPA Glossary of Indoor Air Quality Terms (2011)]

Variance (变动) (biàn dòng)

This is a form of administrative relief that allows property to be used in a way that does not comply with the literal requirements of the zoning ordinance. There are two basic types of variances: use variances and area variances. [Nolon, Well Grounded, p. 456 (ELI 2001)]

Vehicle Miles Traveled (VMT) (车辆行驶英里数) (chē liàng xíng shǐ yīng lǐ shù)

The total number of miles traveled in a given period of time (e.g., day, year) by a given vehicle or fleet of vehicles. VMT, combined with pollution rates per mile traveled, provide an estimate of the total amount of vehicle pollution in a given period of time. [US EPA Glossary of Mobile Source Emissions Terms (2012)]

Ventilation Air (通风换气) (tǒng fēng huàn qì)

Defined as the total air, which is a combination of the air brought inside from outdoors and the air that is being re-circulated within the building. Sometimes, however, used in reference only to the air brought into the system from the outdoors; this document defines this air as "outdoor air ventilation." [US EPA Glossary of Indoor Air Quality Terms (2011)]

Ventilation Rate (通风速率) (tōng fēng sù lǜ)

The rate at which outdoor air enters and leaves a building. Expressed in one of two ways: the number of changes of outdoor air per unit of time (air changes per hour, or "ach") or the rate at which a volume of outdoor air enters per unit of time (cubic feet per minute, or "cfm"). [US EPA Glossary of Indoor Air Quality Terms (2011)]

Vermont Yankee Nuclear Power Corp. v. Natural Resources Defense Council (United States) (佛蒙特州杨基核电站公司诉自然资源保护委员会案 (美国)) ((měi guó) fú méng tè zhōu yáng jī hé diàn zhàn gōng sī sù zì rán zī yuán bǎo hù wěi yuán huì àn)

A 1978 US Supreme Court case holding that, while the Administrative Procedure Act (APA) imposed minimum procedural requirements upon agencies, and agencies may exercise discretion to expand upon those requirements, the APA procedures generally are the only procedures reviewing courts may impose upon the agencies in the absence of such expansion by the agencies.

Vessel (货船) (huò chuán)

Any seagoing vessel or inland waterway craft used for carrying cargo. This restricted use of the term vessel in relation to the transport of radioactive material does not apply in other areas of safety, e.g., a reactor pressure vessel is a vessel as normally understood. [IAEA Safety Glossary, p. 209 (2007)]

Vested Rights (既定权利) (jì dìng quán lì)

[In a land use context] ... are found when a landowner has received approval of a project and has undertaking substantial construction and made substantial expenditures in reliance on that approval. If the landowner's right to develop has vested, it cannot be taken away by a zoning change by the legislature. [Nolon, Well Grounded, p. 456 (ELI 2001)]

Vienna Convention for the Protection Of The Ozone Layer (保护臭氧层维也纳公约) (bǎo hù chòu yǎng céng wéi yě nà gōng yuē)

Ratified by 196 countries, this treaty entered into force in 1985. It aims to provide an international framework for protecting humans and the environment from the destruction of the ozone layer.

Village Board Of Trustees (村管理委员会) (cūn guán lǐ wěi yuán huì)

See Local Legislature.

Volatile Organic Compounds (VOCs) (挥发性有机化合物) (huī fā xìng yǒu jī huà hé wù)

Compounds vaporize and become a gas at room temperature. Common sources which may emit VOCs into indoor air include housekeeping and maintenance products, and building and furnishing materials. In sufficient quantities, VOCs can cause eye, nose, and throat irritations, headaches, dizziness, visual disorders, memory impairment; some are known to cause cancer in animals; some are suspected of causing, or are known to cause, cancer in humans. At present, not much is known about what health effects occur at the levels of VOCs typically found in public and commercial buildings. [US EPA Glossary of Indoor Air Quality Terms (2011)]

W

Warning Point (预警点) (yù jǐng diǎn)

A contact point that is staffed or able to be alerted at all times for promptly responding to, or initiating a response to, an incoming notification (definition (2)), warning message, request for assistance or request for verification of a message, as appropriate, from the IAEA. [IAEA Safety Glossary, p. 211 (2007)]

Warship (军舰) (jūn jiàn)

A ship belonging to the armed forces of a State bearing the external marks distinguishing such ships of its nationality, under the command of an officer duly commissioned by the government of the State and whose name appears in the appropriate service list or its equivalent, and manned by a crew which is under regular armed forces discipline. [Law of the Sea Convention, Art. 29 (1982)]

Waste (废物) (fèi wù)

(1) Substances or objects which are disposed of or are intended to be disposed of or are required to be disposed of by the provisions of national law. [Basel Convention, Art. 2, No. 1 (1989)] (2) Any output from the product or service system which is disposed of. [Von Zharen, ISO 14000, p. 206 (1996)]

Waste Acceptance Requirements (废物接受标准) (fèi wù jiē shōu biāo zhǔn)

Quantitative or qualitative criteria specified by the regulatory body, or specified by an operator and approved by the regulatory body, for radioactive waste to be accepted by the operator of a repository for disposal, or by the operator of a storage facility for storage. Waste acceptance requirements might include, for example, restrictions on the activity concentration or total activity of particular radionuclides (or types of radionuclide) in the waste, or requirements concerning the waste form or packaging of the waste. [IAEA Safety Glossary, p. 212 (2007)]

Waste Form (废物形态) (fèi wù xíng tài)

Waste in its physical and chemical form after treatment and/or conditioning (resulting in a solid product) prior to packaging. The waste form is a component of the waste package. [IAEA Safety Glossary, p. 215 (2007)]

Wastewater (废水) (fèi shuǐ)

Water that has been used and contains dissolved or suspended waste materials. [US EPA Glossary of Climate Change Terms (2011)]

Water Quality Criteria (水质准则) (shuǐ zhì zhǔn zé)

Comprised of numeric and narrative criteria. Numeric criteria are scientifically derived ambient concentrations developed by EPA or states for various pollutants of concern to protect human health and aquatic life. Narrative criteria are statements that describe the desired water quality goal. [US EPA Glossary of NPDES Terms (2004)]

Water Quality Standard (WQS) (水质标准) (shuǐ zhì biāo zhǔn)

A law or regulation that consists of the beneficial use or uses of a waterbody, the numeric and narrative water quality criteria that are necessary to protect the use or uses of that particular waterbody, and an antidegradation statement. [US EPA Glossary of NPDES Terms (2004)]

Water Quality-based Effluent Limit (WQBEL) (流出物水质限度) (liú chū wù shuǐ zhì xiàn dù)

A value determined by selecting the most stringent of the effluent limits calculated using all applicable water quality criteria (e.g., aquatic life, human health, and wildlife) for a specific point source to a specific receiving water for a given pollutant. [US EPA Glossary of NPDES Terms (2004)]

Water Vapor (水蒸气) (shuǐ zhēng qì)

The most abundant greenhouse gas, it is the water present in the atmosphere in gaseous form. Water vapor is an important part of the natural greenhouse effect. While humans are not significantly increasing its concentration, it contributes to the enhanced greenhouse effect because the warming influence of greenhouse gases leads to a positive water vapor feedback. In addition to its role as a natural greenhouse gas, water vapor plays an important role in regulating the temperature of the planet because clouds form when excess water vapor in the atmosphere condenses to form ice and water droplets and precipitation. [US EPA Glossary of Climate Change Terms (2011)]

Waterfowl (水禽) (shuǐ qín)

Birds that ecologically depend on wetlands. [Convention on Wetlands, Art. 1, No. 2 (1987)]

Waters of the United States (美国的水域) (měi guó de shuǐ yù)

All waters that are currently used, were used in the past, or may be susceptible to use in interstate or foreign commerce, including all waters subject to the ebb and flow of the tide. Waters of the United States include all interstate waters and intrastate lakes, rivers, streams (including intermittent streams), mudflats, sand flats, wetlands, sloughs, prairie potholes, wet meadows, playa lakes, or natural ponds. [US EPA Glossary of NPDES Terms (2004)]

Watershed (集水区) (jí shuǐ qū)

(1) In urban areas, a catchment area with an outlet in or affecting a densely populated area. [Urban Soil Primer, p. 72 (2005)] (2) A geographical area within which rainwater and other liquid effluents seep and run into common surface or subsurface water bodies such as streams, rivers, lakes, or aquifers. [Nolon, Well Grounded, p. 456 (ELI 2001)]

Waves (波浪) (bō làng)

A disturbance that travels through space while transferring energy from one point to another. In the ocean context, waves may be the result of wind or geological effects.

Weather (天气) (tiān qì)

Atmospheric condition at any given time or place. It is measured in terms of such things as wind, temperature, humidity, atmospheric pressure, cloudiness, and precipitation. In most places, weather can change from hour-to-hour, day-to-day, and season-to-season. Climate in a narrow sense is usually defined as the "average weather," or more rigorously, as the statistical description in terms of the mean and variability of relevant quantities over a period of time ranging from months to thousands or millions of years. The classical period is 30 years, as defined by the World Meteorological Organization (WMO). These quantities are most often surface variables such as temperature, precipitation, and wind. Climate in a wider sense is the state, including a statistical description, of the climate system. A simple way of remembering the difference is that "climate" is what you expect (e.g., cold winters) and "weather" is what you get (e.g., a blizzard). [US EPA Glossary of Climate Change Terms (2011)]

Wetlands (沼泽地) (zhǎo zé dì)

(1) Areas of marsh, fen, peatland or water, whether natural or artificial, permanent or temporary, with water that is static or flowing, fresh, brackish or salt, including areas of marine water the depth of which at low tide does not exceed six metres. [Convention on Wetlands, Art. 1, No. 1 (1987)] (2) Areas that are inundated or saturated by surface or groundwater at a frequency and duration sufficient to support, and that under normal circumstances do support, a prevalence of vegetation typically adapted for life in saturated soil conditions. Wetlands generally include swamps, marshes, bogs, and similar areas. [US EPA Glossary of NPDES Terms (2004)] (3) [M]ay be either freshwater or tidal. They are typically marked by waterlogged or submerged soils or support a range of vegetation peculiar to wetlands. They provide numerous benefits for human health and property as well as critical habitat for wildlife, and are generally regulated by federal, state, and local laws. [Nolon, Well Grounded, p. 456 (ELI 2001)]

Whales (鲸鱼) (jīng yú)

The common name for a large marine mammal of the order *Cetacea*.

Whole Effluent Toxicity (Wet) (总流出物毒性) (zǒng liú chū wù dú xìng)

The total toxic effect of an effluent measured directly with a toxicity test. [US EPA Glossary of NPDES Terms (2004)]

Wilderness Area: Protected Area Managed Mainly for Wilderness Protection (荒野保护区：主要为管理荒野保护而设立的保护地) (huāng yě bǎo hù qū: zhǔ yào wèi guǎn lǐ huāng yě bǎo hù ér shè lì de bǎo hù dì)

Large area of unmodified or slightly modified land, and/or sea, retaining its natural character and influence, without permanent or significant habitation, which is protected and managed so as to preserve its natural condition. [IUCN, Protected Area Management Categories, (1994)]

Willow (杨柳) (yáng liǔ)

The common name for a species of tree or shrub of the genus *Salix*.

Worker (工作人员) (gōng zuò rén yuán)

Any person who works, whether full time, part time or temporarily, for an employer and who has recognized rights and duties in relation to occupational radiation protection. (A self-employed person is regarded as having the duties of both an employer and a worker.) (From Ref. [1].) [IAEA Safety Glossary, p. 218 (2007)]

Works (土保持技术) (tǔ bǎo chí jì shù)

Any soil conservation technique necessary for the conservation of soil or the mitigation of soil degradation and any operations incidental thereto.

Any "work" should benefit the ecological integrity of soil. [Drafting Legislation for Sustainable Soils: A Guide, p. 94 (2004)]

The World Bank (世界银行) (shì jiè yín háng)

A source of financial and technical assistance to developing countries around the world. [The World Bank (2012)]

The World Charter for Nature (世界自然宪章) (shì jiè zì rán xiàn zhāng)

An agreement adopted in 1982, by United Nations members, proclaiming five guiding principles of conservation: (1) Nature shall be respected and its essential processes shall not be impaired. (2) The genetic viability on the earth shall not be compromised; the population levels of all life forms, wild and domesticated, must be at least sufficient for their survival, and to this end necessary habitats shall be safeguarded. (3) All areas of the earth, both land and sea, shall be subject to these principles of conservation; special protection shall be given to unique areas, to representative samples of all the different types of ecosystems and to the habitats of rare or endangered species. (4) Ecosystems and organisms, as well as the land, marine and atmospheric resources that are utilized by man, shall be managed to achieve and maintain optimum sustainable productivity, but not in such a way as to endanger the integrity of those other ecosystems or species with which they coexist. (5) Nature shall be secured against degradation caused by warfare or other hostile activities.

World Health Organization (WHO) (世界卫生组织) (shì jiè wèi shēng zǔ zhī)

The directing and coordinating authority for health within the United Nations system. It is responsible for providing leadership on global health matters, shaping the health research agenda, setting norms and standards, articulating evidence-based policy options, providing technical support to countries and monitoring and assessing health trends. [World Health Organization (2012)]

World Heritage Fund (世界遗产基金) (shì jiè yí chǎn jī jīn)

An organization that provides funds to support States who are parties to the Convention Concerning the Protection of the World Cultural and Natural Heritage in the protection of sites located in their territories.

World Meteorological Organization (世界气象组织) (shì jiè qì xiàng zǔ zhī)

An agency of the United Nations that is the authoritative voice on the state and behavior of the Earth's atmosphere, its interaction with the oceans, the climate it produces and the resulting distribution of water resources. [World Meteorological Organization (2012)]

World Summit on Sustainable Development, Plan of Implementation (可持续发展世界首脑会议,执行计划) (kě chí xù fā zhǎn shì jiè shǒu nǎo huì yì, zhí xíng jì huà)

A plan of implementation agreed to in Johannesburg in 2002.

World Trade Organization (WTO) (世界贸易组织) (shì jiè mào yì zǔ zhī)

An organization working to open trade and resolve trade problems amongst member states.

World Wildlife Fund (WWF) (世界自然基金会) (shì jiè zì rán jī jīn huì)

An organization working to conserve nature and reduce the most pressing threats to the diversity of life on Earth and to build a future in which people live in harmony with nature. [World Wildlife Fund (2012)]

Z

Zone (中央空调空间) (zhōng yāng kōng tiáo kōng jiān)

The occupied space or group of spaces within a building which has its heating or cooling controlled by a single thermostat. [US EPA Glossary of Indoor Air Quality Terms (2011)]

Zoning Board of Appeals (上诉分区委员会) (shàng sù fēn qū wěi yuán huì)

Under New York State statutes, a zoning board of appeals must be formed when a local legislature adopts its zoning law. It must consist of three to five members. The essential function of the zoning board of appeals is to grant variances. In this capacity, it protects landowners from the unfair application of the laws in particular circumstances. It also hears appeals from the decision of the zoning enforcement officer or building inspector when interpretations of the zoning ordinances are involved. [Nolon, Well Grounded, pp. 456–7 (ELI 2001)]

Zoning District (城市规划分区) (chéng shì guī huà fēn qū)

A part of the community designated by the local zoning law for certain kinds of land uses, such as for single family homes on lots no smaller than one acre or for neighborhood commercial uses. Only the primary permitted land uses, their accessory uses, and any special uses permitted in the zoning district may be placed on the land in that part of the community. [Nolon, Well Grounded, p. 457 (ELI 2001)]

Zoning Enforcement Officer (分区执法官员) (fēn qū zhí fǎ guān yuán)

The local administrative official who is responsible for enforcing and interpreting the zoning law. The local building inspector may be designated as the zoning enforcement officer. Land use applications are submitted to the zoning enforcement officer, who determines whether proposals are in conformance with the use and dimensional requirements of the zoning law. [Nolon, Well Grounded, p. 457 (ELI 2001)]

Zoning Law or Ordinance (分区法律或条例) (fēn qū fǎ lǜ huò tiáo lì)

New York State allows city councils and town boards to adopt zoning regulations by local law or ordinance. Since 1974, village boards of trustees have not had the authority to adopt legislation by ordnance; they may adopt legislation only by local law. Technically, zoning regulations adopted by villages are zoning laws. Only city and town legislatures may adopt zoning ordinances. Zoning regulations, however, are often referred to as zoning ordinances regardless of these technical distinctions. [Nolon, Well Grounded, p. 457 (ELI 2001)]

Zoning Map (区域划分图) (qū yù huà fēn tú)

This map is approved at the time that the local legislature adopts a zoning ordinance. On this map, the zoning district lines are overlaid on a street map of the community. The map divides the community into districts. Each district will carry a designation that refers to the zoning code regulations for that district. By referring to the map, it is possible to identify the use district within which any parcel of land is located. Then by referring to the text of the zoning code, it is possible to discover the uses that are permitted within that district and the dimensional restrictions that apply to building on that land. The zoning map, implemented through the text of the zoning law, constitutes a blueprint of the community over time. [Nolon, Well Grounded, p. 457 (ELI 2001)]

REFERENCES

[Short citation]/Full citation

[Adaptation Fund, About the Adaptation Fund (2011)]
Adaptation Fund, About the Adaptation Fund, http://www.adapta-tion-fund.org/about (accessed May 28, 2012).

[Agreement on the Application of Sanitary and Phytosanitary Measures (2012)]
World Trade Org., Agreement on the Application of Sanitary and Phytosanitary Measures http://www.wto.org/english/docs_e/legal_e/15-sps.pdf (accessed May 28, 2012).

[Agreement on the Network of Aquaculture Centers in Asia and the Pacific (1988)]
Agreement on the Network of Aquaculture Centers in Asia and the Pacific (1988), http://www.ma-law.org.pk/pdflaw/Agreement%20On%20The%20Network%20Of%20Aquaculture%20Centers%20In%20Asia%20And%20the%20Pacific.pdf (accessed May 28, 2012).

[Amazon Cooperation Treaty, July 3, 1978]
Amazon Cooperation Treaty, July 3, 1978, 1202 U.N.T.S. 71.

[Antarctica (1956)]
Antarctica (U.K. v. Arg.), 1956 I.C.J. 12 (Mar. 16).

[ASEAN, About ASEAN (2009)]
About ASEAN, ASEAN, http://www.aseansec.org/about_ASEAN.html (accessed May 28, 2012).

[ASIL Insights]
David P. Fidler, *Weapons of Mass Destruction and International Law*, ASIL Insights (Feb. 2003), http://www.asil.org/insigh97.cfm.

[Australia-Salmon Case (1998)]
WTO Appellate Body Report: Australia–Measures Affecting Importation of Salmon, WT/DS18/AB/R (Oct. 20, 1998), http:// www.wto.org.

[Bali Road Map (2012)]
UNITED NATIONS FRAMEWORK CONVENTION ON CLIMATE CHANGE, BALI ROAD MAP, http://unfccc.int/key_documents/bali_road_map/items/6447.php (accessed May 28, 2012).

[Basel Convention (1989)]
Basel Convention on the Control of the Transboundary Movements of Hazardous Wastes and Their Disposal, Mar. 22, 1989, 1673 U.N.T.S. 125, http://www.basel.int/text/con-e-rev.pdf.

[Basel Protocol (1999)]
Protocol on Liability and Comp. for Damage Resulting from

Transboundary Movements of Hazardous Wastes and Their Disposal art. 1, Dec. 10, 1999 U.N. Doc. UNEP/CHW.1/WG/1/9/2.

[Calvert Cliffs' (1971)]
Calvert Cliffs' Coordinating Comm., Inc. v. U. S. Atomic Energy Comm'n, 449 F.2d 1109 (D.C. Cir. 1971).

[CIESIN]
International Convention Relating to Intervention on the High Seas in Cases of Oil (1970) 9 I.L.M. 25, http://sedac.ciesin.org/entri/register/reg-052.rrr.html (accessed May 28, 2012).

[CIESIN Civil Liability]
International Convention on Civil Liability for Oil Pollution Damage, Nov. 29, 1969, 9 I.L.M. 45, http://sedac.ciesin.columbia.edu/entri/texts/civil.liability.oil.pollution.damage.1969.html (accessed May 28, 2012).

[CIESIN Oil Pollution at Sea]
International Convention for the Prevention of Pollution of the Sea by Oil, 12 U.S.T. 2989, 327 U.N.T.S. 3 (1954), http://sedac.ciesin.org/entri/texts/pollution.of.sea.by.oil.1954.html (accessed May 28, 2012).

[C.K. Wentworth, Clastic Sediments (1922)]
C.K. Wentworth, *A Scale of Grade and Class Terms for Clastic Sediments*, 30, J. Geol. 377 (1922).

[Clean Development Mechanism]
UNFCCC, *About CDM*, https://cdm.unfccc.int/about/index.html (accessed November 29, 2012).

[Clean Production Action]
Clean Production Action, http://www.cleanproduction.org/Steps.Process.UN.php (accessed May 28, 2012).

[Clean Water Act, 33 U.S.C. § 1251 *et seq.* (1972)]
Clean Water Act, 33 U.S.C. § 1251 *et seq.* (1972).

[Convention on Assistance in the Case of a Nuclear Accident (1987)]
Convention on Assistance in the Case of a Nuclear Accident or Radiological Emergency, Sept. 26, 1986, IAEA Doc. INFCIRC/336, 25 I.L.M. 1377 (entered into force Feb. 26, 1987), http://www.iaea.org/Publications/Documents/Infcircs/Others/infcirc336.shtml.

[Convention on Biological Diversity (1992)]
Convention on Biological Diversity, June 5, 1992, 1760 U.N.T.S. 79, 143; 31 I.L.M. 818 (1992).

[Convention on Civil Defence Assistance (2000)]
Framework Convention on Civil Defense Assistance, May 22, 2000, 2172 U.N.T.S. 231.

[Convention on Marine Pollution (1974)]
Convention for the Prevention of Marine Pollution from Land-Based Sources, June 4, 1974, 1546 U.N.T.S. 119; 13 I.L.M. 352 (1974).

[Convention on Nuclear Safety]
 Convention On Nuclear Safety, IAEA, Sept. 20, 1994, IAEA Doc.
 INFCIRC/449, 33 I.L.M. 1514.
[Convention on the Prevention of Marine Pollution by Dumping of
 Wastes and Other Matter]
 Convention on the Prevention of Marine Pollution by Dumping of
 Wastes and Other Matter, Dec. 29, 1972, 26 U.S.T. 2403, 1046 U.N.T.S.
 138 (entered into force 2006).
[Convention on Wetlands]
 Convention on Wetlands of International Importance especially as
 Waterfowl Habitat, Feb. 2, 1971, 996 U.N.T.S. 245, 111 I.L.M. 963
 (entered into force Dec. 21, 1975).
[Convention to Combat Desertification (1994)]
 United Nations Convention to Combat Desertification in Countries
 Experiencing Serious Drought and/or Desertification, Particularly in
 Africa, June 17, 1994, 1954 U.N.T.S. 3.
[Corfu Channel (1947)]
 Corfu Channel (U.K. & Ir. v. Alb.), 1949 I.C.J. Pleadings 8 (May 22,
 1947).
[Decision 1 (2011) - ATCM XXXIV - CEP XIV, Buenos Aires]
 Agreed Measures for the Conservation of Antarctic Fauna and Flora,
 June 13, 1964, 17 U.S.T. 991 (entered into force Sept. 1, 1976), http://
 www.ats.aq/documents/recatt/att080_e.pdf (accessed May 28, 2012).
[Declaration on the Human Environment]
 UNEP, *Declaration of the United Nations Conference on the Human
 Environment*, http://www.unep.org/Documents.Multilingual/Default.
 asp?documentid=97&articleid=1503 (accessed November 29, 2012).
[Deepwater Report to the President (2011)]
 Nat'l Comm'n on the BP Deepwater Horizon Oil Spill and Offshore
 Drilling, Deepwater: The Gulf Oil Disaster and the Future of
 Offshore Drilling (report to the President) (2011), http://www.oil
 spillcommission . gov / sites / default / files / documents / DEEPWATER _
 ReporttothePresident_FINAL.pdf (accessed May 28, 2012).
[Drafting Legislation for Sustainable Soils: A Guide (2004)]
 IAN HANNAM & BEN BOER, DRAFTING LEGISLATION FOR SUSTAINABLE
 SOILS: A GUIDE, IUCN Envtl. Law Programme (2004), http://data.iucn.
 org/dbtw-wpd/edocs/EPLP-052.pdf.
[ECOSOC (2012)]
 UNITED NATIONS ECON. AND SOCIAL COUNCIL, http://www.un.org/en/
 ecosoc/about/index.shtml (accessed May 28, 2012).
[Encyclopedia of Earth]
 Vito De Lucia, *Common But Differentiated Responsibility*,

ENCYCLOPEDIA OF EARTH (January 27, 2007), http://www.eoearth.org/article/Common_but_differentiated_responsibility?topic=49477.

[EPA Pollution Prevention]
EPA, Pollution Prevention, http://www.epa.gov/ems/ (accessed November 29, 2012).

[European Council (2012)]
EUROPEAN COUNCIL, http://www.european-council.europa.eu/the-institution?lang=en (accessed May 28, 2012).

[FAO (2012)]
United Nations Food and Agric. Org. of the United Nations, http://www.fao.org/about/en/ (accessed May 28, 2012).
Freedom of Information Act, 5 U.S.C. § 552 (1966)]
Freedom of Information Act, 5 U.S.C. § 552 (1966).

[French Environmental Code]
CODE DE L'ENVIRONNEMENT / ENVIRONMENTAL CODE, http://195.83.177.9/upl/pdf/code_40.pdf (unofficial English translation).

[Gabcikovo-Nagymaros Project (1997)]
Case Concerning the Gabcíkovo-Nagymaros Project (Hung. v. Slovk.) 1997 I.C.J. 92 (Sept. 25), http://www.icj-cij.org/docket/files/92/7375.pdf (accessed May 28, 2012).

[GEF, About GEF (2011)]
GEF, About GEF, http://www.thegef.org/gef/whatisgef (accessed May 28, 2012).

[GEF, Country Eligibility (2012)]
GEF, County Eligibility, http://www.thegef.org/gef/node/1432 (accessed May 28, 2012).

[GEF, GEF Council (2012)]
GEF, GEF Council, http://www.thegef.org/gef/council (accessed May 28, 2012).

[GEF, GEF-Administered Trust Funds (2012)]
GEF, GEF-Administered Trust Funds, http://www.thegef.org/gef/node/2042 (accessed May 28, 2012).

[GEMET Thesaurus (2012)]
EIONET GEMET THESAURUS, http://www.eionet.europa.eu/gemet/concept?cp=1024 (accessed May 28, 2012).

[Germany Federal Environment Agency]
UMWELTBUNDESAMT (GERMANY FEDERAL ENVIRONMENT AGENCY), http://www.umweltbundesamt.de/umweltrecht-e/umweltgesetzbuch.htm.

[Global Crop Diversity Trust, About Us (2012)]
Global Crop Diversity Trust, About Us, http://www.croptrust.org/main/laboutus.php (accessed May 28, 2012).

[Global Crop Diversity Trust, Founders (2012)]
    Global Crop Diversity Trust, Founders, http://www.croptrust.org/
    main/content/founders (accessed May 28, 2012).
[Glossary of Terms for Negotiators of MEAs (2007)]
    UNEP, GLOSSARY OF TERMS FOR NEGOTIATORS OF MULTILATERAL
    ENVTL. AGREEMENTS (2007).
[Greenpeace, About Greenpeace (2012)]
    Greenpeace, About Greenpeace, http://www.greenpeace.org/interna-
    tional/en/about/ (accessed May 28, 2012).
[IUCN, Protected Area Management Categories (1994)]
    IUCN, GUIDELINES FOR PROTECTED AREA MANAGEMENT CATEGORIES,
    IUCN, Gland, Switzerland and Cambridge, UK (1994).
[IAEA Safety Glossary (2007)]
    INTERNATIONAL ATOMIC ENERGY AGENCY, IAEA SAFETY GLOSSARY
    TERMINOLOGY USED IN NUCLEAR SAFETY AND RADIATION PROTECTION
    (2007).
[ICETT, Itai-itai Disease (2010)]
    INT'L CENTER FOR ENVTL. TECH. TRANSFER, ITAI-ITAI DISEASE, http://
    www.icett.or.jp/english/abatement/toyama/disease.html (accessed May
    28, 2012).
[ICJ, Nuclear Tests (New Zealand v. France) (1974)]
    ICJ, Nuclear Test Case (New Zealand v. France) Summary 88–89
    (1974), http://www.icj-cij.org/docket/files/59/6117.pdf (accessed May
    28, 2012).
[Int'l Atomic Energy Agency (2012)]
    Int'l Atomic Energy Agency, http://www.iaea.org/About/about-iaea.
    html (accessed May 28, 2012).
[Int'l Comm'n for the Conservation of Atlantic Tunas (2012)]
    Int'l Comm'n for the Conservation of Atlantic Tunas, http://www.iccat.
    es/en/ (accessed May 28, 2012).
[Int'l Conv. for the Prot. of New Varieties of Plants]
    International Convention for the Protection of New Varieties of Plants:
    Act of 1961/1972, 33 UST. 2703, 1861 U.N.T.S. 281, http://www.wipo.
    int/wipolex/en/other_treaties/text.jsp?doc_id=131052&file_id=193302
    (accessed May 28, 2012).
[Int'l Court of Justice (2012)]
    Int'l Court of Justice, http://www.icj-cij.org/information/index.
    php?p1=7&p2=2 (accessed May 28, 2012).
[Int'l Joint Comm'n (2012)]
    Int'l Joint Comm'n: About the Great Lakes Water Quality Agreement,
    http://www.ijc.org/rel/agree/quality.html (accessed May 28, 2012).
[Int'l Labor Org. (2012)]

Int'l Labor Org., http://www.ilo.org/global/about-the-ilo/lang--en/index.htm (accessed May 28, 2012).

[Int'l Law Association (2004)]
JOSEPH W. DELLAPENNA, INT'L LAW ASSOCIATION, BERLIN RULES ON WATER RESOURCES (2004).

[Int'l Maritime Org. (2012)]
Int'l Maritime Org., http://www.imo.org/About/Pages/Default.aspx (accessed May 28, 2012).

[Int'l Org. for Standardization (2012)]
Int'l Org. for Standardization, http://www.iso.org/iso/about.htm (accessed May 28, 2012).

[Int'l Renewable Energy Agency (2012)]
Int'l Renewable Energy Agency, http://www.irena.org/Menu/index.aspx?PriMenuID=13&mnu=Pri (accessed May 28, 2012).

[Int'l Union for Conservation of Nature (2012)]
Int'l Union for the Conservation of Nature, About IUCN, http://iucn.org/about/ (accessed November 29, 2012).

[IPCC, CLIMATE CHANGE 2001: WORKING GROUP III: MITIGATION (2001)]
IPCC, Climate Change 2001: Working Group III: Mitigation, Glossary (2001), http://www.ipcc.ch/ipccreports/tar/wg3/454.htm.

[IPPC Directive (2012)]
European Comm'n, IPPC Directive, http://ec.europa.eu/environment/air/pollutants/stationary/ippc/index.htm (accessed May 28, 2012).

[ITTO]
Int'l Tropical Timber Org., http://www.itto.int/about_itto/ (accessed May 28, 2012).

[Japan Envt'l Health Comm. Report (1999)]
ENVTL. HEALTH COMM. OF THE CENTRAL ENV'T COUNCIL, ENV'T AGENCY & LIVING ENV'T COUNCIL & FOOD SANITATION INVESTIGATION COUNCIL, MINISTRY OF HEALTH & WELFARE, REPORT ON TOLERABLE DAILY INTAKE (TDI) OF DIOXINS AND RELATED COMPOUNDS (Japan) (June 1999), http://www.env.go.jp/en/chemi/dioxins/tdi_report.pdf (accessed May 28, 2012).

[Japan Ministry of the Environment, Minimata Disease (2002)]
JAPAN MINISTRY OF THE ENV'T, MINIMATA DISEASE: THE HISTORY & MEASURES (2002), http://www.env.go.jp/en/chemi/hs/minamata2002/ (accessed May 28, 2012).

[Land & Environment Court, About Us (2011)]
LAND & ENV'T COURT, ABOUT US, http://www.lawlink.nsw.gov.au/lawlink/lec/ll_lec.nsf/pages/LEC_aboutus (accessed May 28, 2012).

[Law of the Sea Convention (1982)]
United Nations Convention on the Law of the Sea, Dec. 10, 1982, 1833

U.N.T.S. 3, 397; 21 I.L.M. 1261 (1982) (Common name, Law of the Sea Convention).

[Mass. v. E.P.A., 549 U.S. at 559]

Massachusetts v. E.P.A., 549 U.S. 497 (2007).

[Montreal Protocol (1987)]

Montreal Protocol on Substances that Deplete the Ozone Layer, Sept. 16, 1987, 1522 U.N.T.S. 3 (entered into force Jan. 1, 1989).

[Multilateral Environmental Agreement Negotiator's Handbook (2007)]

MULTILATERAL ENVTL. AGREEMENT NEGOTIATOR'S HANDBOOK (2007), http://unfccc.int/resource/docs/publications/negotiators_handbook.pdf (accessed May 28, 2012).

[NASA Earth Observatory (2012)]

NASA Earth Observatory, http://earthobservatory.nasa.gov/Glossary/index.php?mode=all (accessed May 28, 2012).

[NASA, Toxic Sludge in Hungary (2012)]

NASA, Toxic Sludge in Hungary, http://earthobservatory.nasa.gov/NaturalHazards/view.php?id=46360 (accessed May 28, 2012).

[National Academies Press]

THE NATIONAL ACADEMIES PRESS, ON THE FULL AND OPEN EXCHANGE OF SCIENTIFIC INFORMATION, http://www.nap.edu/readingroom.php?book=exch&page=attach1.html (accessed May 28, 2012).

[NOAA Gulf of Mexico Hypoxia Assessment]

NOAA, National Centers for Coastal Ocean Science Gulf of Mexico Hypoxia Assessment (2000), http://oceanservice.noaa.gov/products/pubs_hypox.html.

[NOAA, The Monsoon (2012)]

NOAA NAT'L WEATHER SERVICE FORECAST OFFICE, THE MONSOON, http://www.wrh.noaa.gov/fgz/science/monsoon.php?wfo=fgz (accessed May 28, 2012).

[NOAA, Terms Used By Meteorologists (2012)]

NOAA, TERMS USED BY METEOROLOGISTS, FORECASTERS, WEATHER OBSERVERS, & IN WEATHER FORECASTS, http://www.erh.noaa.gov/er/box/glossary.htm (accessed May 28, 2012).

[Nolon, Well Grounded (ELI 2001)]

JOHN R. NOLON, WELL GROUNDED: USING LOCAL LAND USE AUTHORITY TO ACHIEVE SMART GROWTH (Envtl. Law Inst., 445–57 (2001)).

[OAU (2012)]

Republic of South Africa, Organization of African Union, http://www.dfa.gov.za/foreign/Multilateral/africa/oau.htm, (accessed May 28, 2012).

[OPRC 1990]

International Convention on Oil Pollution Preparedness, Response

and Co-operation, Nov. 30, 1990, 30 I.L.M. 735, http://www.ecolex.
org/ecolex/ledge/view/RecordDetails?id=TRE-001109&index=treaties
(accessed May 28, 2012).

[Org. for Econ. Cooperation & Dev. (2012)]
Org. for Econ. Cooperation & Dev., http://www.oecd.org/pages/
0,3417,en_36734052_36734103_1_1_1_1_1,00.html (accessed May 28,
2012).

[OSHA (2012)]
US Dep't of Labor Occupational Safety And Health Administration,
http://www.osha.gov/about.html (accessed May 28, 2012).

[OSHA Glossary (2012)]
US DEP'T OF LABOR OCCUPATIONAL SAFETY AND HEALTH ADMINI-
STRATION GLOSSARY, http://www.dol.gov/elaws/osha/lead/glossary.asp
(accessed May 28, 2012).

[OT Ministry of Municipal Affairs & Housing Land Use Planning
Definitions (2007)]
ONTARIO MINISTRY OF MUNICIPAL AFFAIRS AND HOUSING, LAND USE
PLANNING DEFINITIONS, http://www.mah.gov.on.ca/Page1491.aspx
(accessed May 28, 2012).

[Report of the Japanese Government to IAEA (2011)]
REPORT OF JAPANESE GOV'T TO THE IAEA MINISTERIAL CONFERENCE ON
NUCLEAR SAFETY: THE ACCIDENT AT TEPCO'S FUKUSHIMA NUCLEAR
POWER STATIONS (June 2011), http://www.kantei.go.jp/foreign/kan/
topics/201106/iaea_houkokusho_e.html (accessed May 28, 2012).

[Restrictions on Imports of Tuna (1994)]
Report of the Panel, United States: Restrictions on Imports of Tuna,
(June 16, 1994), GATT Panel Report DS29/R, http://sul-derivatives.
stanford.edu/derivative?CSNID=91790155&mediaType=application/
pdf (accessed May 28, 2012).

[Rotterdam Convention (1998)]
Rotterdam Convention on the Prior Informed Consent Procedure for
Certain Hazardous Chemicals and Pesticides in Int'l Trade, Sept. 10,
1998, 38 I.L.M. 1.

[Stockholm Convention (2001)]
Stockholm Convention on Persistent Organic Pollutants, May 22, 2001,
S. Treaty Doc. No. 107-5, U.N. Doc. UNEP/POPS/CONF/2, 40 I.L.M.
532.

[The China Quarterly (2012)]
Goubin Yang, *Environmental NGOs and Institutional Dynamics in
China*, THE CHINA QUARTERLY, http://bc.barnard.columbia.
edu/~gyang/Yang_ENGOs.pdf (accessed May 28, 2012).

[The World Bank (2012)]

The World Bank, http://web.worldbank.org/WBSITE/EXTERNAL/
EXTABOUTUS/0,,contentMDK:20103838~menuPK:1696997~page
PK:51123644~piPK:329829~theSitePK:29708,00.html (accessed May
28, 2012).

[Trail Smelter Case (1941)]
Trail Smelter Case (US v. Can.), 3 R.I.A.A 1905 (Trail Smelter Arb.
Trib. 1938 & 1941), http://untreaty.un.org/cod/riaa/cases/vol_III/1905-
1982.pdf.

[TVA v. Hill (1978)]
Tenn. Valley Auth. v. Hill, 437 U.S. 153 (1978).

[Urban Soil Primer (2005)]
SCHEYER, J.M., & K.W. HIPPLE, URBAN SOIL PRIMER. UNITED STATES
DEP'T OF AGRICULTURE, NATURAL RES. CONSERVATION SERVICE (2005),
http://soils.usda.gov/use (accessed May 28, 2012).

[U.N. Agenda 21 (1992)]
U.N. Conference on Env't and Dev., Agenda 21, U.N. Doc. A/
CONF.151/4 (Parts I-IV) (1992).

[U.N. Glossary of Env't Statistics (1997)]
UNITED NATIONS, GLOSSARY OF ENV'T STATISTICS, STUDIES IN METHODS,
Series F, No. 67 (1996).

[UNDP (2012)]
United Nations Dev. Programme, http://www.undp.org/content/undp/
en/home/operations/about_us.html, (accessed May 28, 2012).

[UNEP (2012)]
United Nations Env't Programme, About UNEP, http://www.unep. org/
Documents.Multilingual/Default.asp?DocumentID=43, (accessed May
28, 2012).

[UNEP Green Economy Report (2011)]
UNEP, TOWARDS A GREEN ECONOMY: PATHWAYS TO SUSTAINABLE
DEVELOPMENT AND POVERTY ERADICATION (2011), available at http://
www.unep.org/greeneconomy/greeneconomyreport/tabid/29846/
default.aspx.

[UNEP Guide to Climate Neutrality (2008)]
UNEP, Kick the Habit, A UN Guide to Climate Neutrality (2008).

[UNEP Judicial Handbook (2005)]
DINAH SHELTON & ALEXANDRE KISS, UNITED NATIONS ENVIRONMENT
PROGRAM, JUDICIAL HANDBOOK ON ENVIRONMENTAL LAW (UNEP
2005).

[UNEP, Management of Industrial Accident Prevention & Preparedness
(1996)]
UNEP, MGMT. OF INDUS. ACCIDENT PREVENTION & PREPAREDNESS
(1996).

[UNEP, Protocol to the Antarctic Treaty on Environmental Protection (1998)]
UNEP, PROTOCOL TO THE ANTARCTIC TREATY ON ENVTL. PROTECTION (1998).

[UNESCO (2012)]
United Nations Educ., Scientific and Cultural Org., http://www.unesco.org/new/en/unesco/about-us/who-we-are/introducing-unesco/, (accessed May 28, 2012).

[UNESCO, The Precautionary Principle (2005)]
UNESCO, THE PRECAUTIONARY PRINCIPLE (2005).

[UNFCCC, The Mechanisms Under the Kyoto Protocol]
UNFCCC, The Mechanisms Under the Kyoto Protocol, http://unfccc.int/kyoto_protocol/mechanisms/items/1673.php (accessed May 28, 2012).

[United Nations, About the General Assembly (2012)]
United Nations: About the General Assembly, http://www.un.org/en/ga/ (accessed May 28, 2012).

[United Nations Framework Convention on Climate Change (1992)]
United Nations Framework Convention on Climate Change, May 9, 1992, S. Treaty Doc No. 102-38, 1771 U.N.T.S. 107, http://unfccc.int/essential_background/convention/items/2627.php.

[United Nations, Trusteeship Council (2012)]
United Nations, Trusteeship Council, http://www.un.org/en/mainbodies/trusteeship/ (accessed May 28, 2012).

[United States – Taxes on Automobiles (1994)]
Report of the Panel, United States – Taxes on Automobiles, (Oct. 11, 1994), DS31/R, 11.

[UNU IAS Pocket Guide (2007)]
BALAKRISHNA PISUPATI, UNITED NATIONS UNIVERSITY, UNU IAS POCKET GUIDE ACCESS TO GENETIC RES., BENEFIT SHARING AND BIOPROSPECTING (UNU 2007).

[UNOOSA]
UNOOSA, *Convention on International Liability for Damage Caused by Space Objects*, http://www.oosa.unvienna.org/oosa/SpaceLaw/liability.html (accessed November 29, 2012).

[USDA-NRCS Understanding Soil Risks and Hazards (2004)]
USDA-NRCS, UNDERSTANDING SOIL RISKS AND HAZARDS: USING SOIL SURVEY TO IDENTIFYING AREAS WITH RISKS AND HAZARDS TO HUMAN LIFE AND PROPERTY (Muckel, ed. 2004), available at http://soils.usda.gov/use/risks.html.

[US DOE, Geothermal Energy & the Environment (2007)]

US DEP'T OF ENERGY, A GUIDE TO GEOTHERMAL ENERGY & THE ENV'T (2007), http://geo-energy.org/reports/environmental%20guide. pdf (accessed May 28, 2012).

[US DOJ, The Antarctic Treaty (2001)]
The Antarctic Treaty, Dec. 1, 1959, 402 U.N.T.S. 71 (entered into force June 23, 1961), http://www.state.gov/www/global/arms/treaties/arctic1. html.

[US EPA, Acetic Acid Factsheet (2001)]
US EPA: Acetic Acid Factsheet, issued Mar. 2001, http://www.epa.gov/ oppbppd1/biopesticides/ingredients/factsheets/factsheet_044001.htm (accessed May 28, 2012).

[US EPA, Arsenic Compounds (2007)]
US EPA, Arsenic Compounds, updated Nov. 6, 2007, http://www.epa. gov/ttn/atw/hlthef/arsenic.html (accessed May 28, 2012).

[US EPA, Arsenic in Drinking Water (2011)]
US EPA, Arsenic in Drinking Water, updated Dec. 23, 2011, http:// water.epa.gov/lawsregs/rulesregs/sdwa/arsenic/index.cfm (accessed May 28, 2012).

[US EPA, CERCLA (2012)]
US Envtl Prot. Agency: CERCLA Overview, updated Dec. 12, 2011, http://www.epa.gov/superfund/policy/cercla.htm (accessed May 28, 2012).

[US EPA, Clean Air Act (2012)]
US Envtl Prot. Agency: Clean Air Act, updated Feb. 17, 2012, http:// epa.gov/air/caa/caa_history.html (accessed May 28, 2012).

[US EPA Envtl. Justice (2012)]
US Envtl. Prot. Agency: Envtl. Justice, updated Feb. 27, 2012, http:// www.epa.gov/environmentaljustice/ (accessed May 28, 2012).

[US EPA, EPCRA (2012)]
US Envtl. Prot. Agency, Emergency Planning and Cmty. Right-To-Know Act, updated Feb. 14, 2012, http://www.epa.gov/agriculture/lcra. html (accessed May 28, 2012).

[US EPA Glossary of Climate Change Terms (2011)]
US Envtl. Prot. Agency, Glossary of Climate Change Terms, updated Nov. 2, 2011, http://www.epa.gov/climatechange/glossary.html (accessed May 28, 2012).

[US EPA Glossary of Indoor Air Quality Terms (2011)]
US ENVTL. PROT. AGENCY, GLOSSARY OF INDOOR AIR QUALITY TERMS, updated Nov. 2, 2011, http://www.epa.gov/iaq/glossary.html (accessed May 28, 2012).

[US EPA Glossary of Mobile Source Emissions Terms (2012)]
US ENVTL. PROT. AGENCY, GLOSSARY OF MOBILE SOURCE EMISSIONS,

updated Jan. 3, 2012, http://www.epa.gov/otaq/invntory/overview/definitions.htm (accessed May 28, 2012).

[US EPA Glossary of NPDES Terms (2004)]
US ENVTL. PROT. AGENCY, GLOSSARY OF NAT'L POLLUTANT DISCHARGE ELIMINATION SYS., updated Mar. 23, 2004, http://cfpub.epa.gov/npdes/glossary.cfm (accessed May 28, 2012).

[US EPA, Hydroelectricity (2007)]
US Envtl. Prot. Agency, Hydroelectricity, updated Dec. 28, 2007, http://www.epa.gov/cleanenergy/energy-and-you/affect/hydro.html (accessed May 28, 2012).

[US EPA, IRIS Glossary (2011)]
US ENVTL. PROT. AGENCY, IRIS GLOSSARY, updated Apr. 11, 2011, http://www.epa.gov/iris/gloss8_arch.htm (accessed May 28, 2012).

[US EPA, NEPA (2012)]
NEPA Sec. 102, 42 USC § 4332, http://www.epa.gov/region2/spmm/r2nepa.htm (accessed May 28, 2012).

[US EPA, OAR Policy Guidance Fact Sheet (2011)]
US Envtl. Prot. Agency, TTN OAR Policy & Guidance Fact Sheet, updated Apr. 19, 2011, http://www.epa.gov/ttn/caaa/t3/fact_sheets/forcemajproposalfs.html (accessed May 28, 2012).

[US EPA Pesticides (2012)]
US Envtl. Prot. Agency: Pesticides, updated Feb. 27, 2012, http://www.epa.gov/pesticides/about/index.htm (accessed May 28, 2012).

[US EPA, RCRA (2012)]
US Envtl. Prot. Agency, Summary of the Resource Conservation and Recovery Act, updated Feb. 24, 2012, http://www.epa.gov/lawsregs/laws/rcra.html (accessed May 28, 2012).

[US EPA, TSCA (2012)]
US Envtl. Prot. Agency, Summary of Toxics Substances Control Act, updated Feb. 24, 2012, http://www.epa.gov/lawsregs/laws/tsca.html (accessed May 28, 2012).

[US EPA, When & How to Use ADR (2012)]
US ENVTL. PROT. AGENCY, WHEN & HOW TO USE ADR, updated Feb. 6, 2012, http://www.epa.gov/region1/enforcement/adr/use.html (accessed May 28, 2012).

[US Fish & Wildlife Service, Endangered Species Glossary (2012)]
US FISH & WILDLIFE SERVICE, ENDANGERED SPECIES GLOSSARY (2011), http://www.fws.gov/midwest/endangered/glossary/index.html (accessed May 28, 2012).

[US GAO, Environmental Liabilities (2005)]
US GEN'L ACCOUNTING OFFICE, ENVTL. LIABILITIES, EPA SHOULD DO MORE TO ENSURE THAT LIABLE PARTIES MEET THEIR CLEANUP

OBLIGATIONS, GAO-05-658 (Aug. 2005), http://www.gao.gov/new.
items/d05658.pdf (accessed May 28, 2012).
[USGS Circular (2000)]
W. BRIAN HUGHES ET AL., WATER QUALITY IN THE SANTEE RIVER BASIN
AND COSTAL DRAINAGES, NORTH AND SOUTH, 1995–98, US GEOLOGICAL
SURVEY CIRCULAR 1206 (2000), http://pubs.water.usgs.gov/circ1206/
(accessed May 28, 2012).
[USGS, (2012)]
US GEOLOGICAL SERVICE, FOSSILS, http://www.usgs.gov/science/science.
php?term=414 (accessed May 28, 2012).
[USGS, Hydrology Primer (2011)]
US GEOLOGICAL SERVICE, HYDROLOGY PRIMER, updated Dec. 22,
2011, http://ga.water.usgs.gov/edu/hydrology.html (accessed May 28,
2012).
[USGS, Mineral Resources Online Spatial Data (2011)]
US GEOLOGICAL SERVICE, MINERAL RES. ONLINE SPATIAL DATA,
updated Dec. 8, 2011, http://tin.er.usgs.gov/geology/state/sgmc-lith.
php?text=loess (accessed May 28, 2012).
[USGS, Science in Your Watershed (2011)]
US GEOLOGICAL SERVICE, SCIENCE IN YOUR WATERSHED, GEN.
INTRODUCTION & HYDROLOGIC DEFINITIONS (2011), http://water.usgs.
gov/wsc/glossary.html (accessed May 28, 2012).
[USGS, Water Science Glossary of Terms (2011)]
US GEOLOGICAL SERVICE, WATER SCIENCE GLOSSARY OF TERMS, updated
Dec. 22, 2011, http://ga.water.usgs.gov/edu/dictionary.html (accessed
May 28, 2012).
[von Zharen, ISO 14000 (1996)]
W.M. VON ZHAREN, ISO 14000 UNDERSTANDING THE ENVTL. STAN-
DARDS (Gov't Institutes, Inc., Rockville, MD (1996)).
[What is Carbon Trading?]
Oklahoma Forestry Service, *What is Carbon Trading?* http://www.for-
estry.ok.gov/what-is-carbon-trading (accessed November 29, 2012).
[World Bank, Force Majeure Clauses (2012)]
WORLD BANK, FORCE MAJEURE CLAUSES: CHECKLIST AND SAMPLE
WORDING, http://siteresources.worldbank.org/INTINFANDLAW/
Resources/Forcemajeurechecklist.pdf (accessed May 28, 2012).
[World Health Organization (2012)]
WORLD HEALTH ORG., http://www.who.int/about/en/ (accessed May 28,
2012).
[World Meteorological Organization (2012)]
World Meteorological Org., http://www.wmo.int/pages/about/index_
en.html (accessed May 28, 2012).

[World Wildlife Fund (2012)]
    World Wildlife Fund, http://www.worldwildlife.org/who/index.html
    (accessed May 28, 2012).